LECTURES ON EXPONENTIAL DECAY OF SOLUTIONS OF SECOND-ORDER ELLIPTIC EQUATIONS: BOUNDS ON EIGENFUNCTIONS OF N-BODY SCHRÖDINGER OPERATORS

by

Shmuel Agmon

T0329793

Mathematical Notes 29

Princeton University Press

and

University of Tokyo Press

1982

Copyright © 1982 by Princeton University Press

All Rights Reserved

Published in Japan exclusively
by University of Tokyo Press;
in other parts of the world by
Princeton University Press

Printed in the United States of America by
Princeton University Press, 41 William Street,
Princeton, New Jersey 08540

The Princeton Mathematical Notes are edited by William Browder,
Robert Langlands, John Milnor, and Elias M. Stein

Library of Congress Cataloging in Publication Data
will be found on the last printed page of this book

Table of Contents

Introduction		3
Chapter 0	Preliminaries	8
Chapter 1	The Main Theorem	11
Chapter 2	Geometric Spectral Analysis	32
Chapter 3	Self- Adjointness	41
Chapter 4	L^2 Exponential Decay. Applications to eigenfunctions of N-body Schrödinger operators	52
Chapter 5	Pointwise Exponential Bounds	83
Appendix 1	Approximation of Metrics and Completeness	99
Appendix 2	Proof of Lemma 1.2	102
Appendix 3	Proof of Lemma 2.2	104
Appendix 4	Proof of Lemma 5.7	110
Bibliographical Comments		112
References		115

Introduction

This volume presents an edited and slightly revised version of notes of lectures given at the University of Virginia in the fall of 1980. The subject of these lectures is the phenomenon of exponential decay of solutions of second order elliptic equations in unbounded domains. By way of introduction we discuss briefly the special problem of exponential decay of eigenfunctions of Schrödinger operators, a problem which motivated the present investigations.

Consider the Schrödinger differential operator $P = -\Delta + V(x)$ on $\mathbb{R}^n$ where V is a real function in $L^1_{loc}(\mathbb{R}^n)$. Assume also that $V_- \in L^p_{loc}(\mathbb{R}^n)$ for some $p > n/2$ and that

$$\inf\{(P\varphi,\varphi) : \varphi \in C_0^\infty(\mathbb{R}^n),\ \|\varphi\|_{L^2(\mathbb{R}^n)} = 1\} > -\infty$$

where

$$(P\varphi,\varphi) = \int_{\mathbb{R}^n} (|\nabla\varphi|^2 + V|\varphi|^2)dx.$$

Under these conditions P admits a unique self-adjoint realization in $L^2(\mathbb{R}^n)$ which we denote by H (see Chapter 3). The essential spectrum of H (which is bounded from below) is denoted by $\sigma_{ess}(H)$. We set

$$\Sigma = \inf \sigma_{ess}(H),$$

$\Sigma = +\infty$ if $\sigma_{ess}(H)$ is empty.

There is a general decay phenomenon of eigenfunctions of H. Namely, for a general class of potentials V any eigenfunction of H with eigenvalue in the discrete spectrum decays exponentially. This phenomenon was studied extensively in the literature for the special class of eigenfunctions with eigenvalues situated *below* the bottom of the essential spectrum Σ. Thus it follows from the works of O'Conner [31] Combes and Thomas [7] and Simon [37], under some restrictions on V, that if $\psi(x)$ is an eigenfunction of H with eigenvalue $\mu < \Sigma$, then

$$|\psi(x)| \le C_\alpha e^{-\alpha|x|} \quad on \quad \mathbb{R}^n \tag{1}$$

where α is any number such that $\alpha < (\Sigma-\mu)^{1/2}$ and C_α is some constant This type of estimate for ψ, which may be referred to as the *isotropic* estimate, is precise if $V(x) \to 0$ as $|x| \to \infty$ However, if $V(x)$ does not tend to the same limit in all directions as $|x| \to \infty$ one would expect that the isotropic estimate could be replaced by a more precise non-isotropic estimate of the form

$$|\psi(x)| \leq C_\varepsilon e^{-(1-\varepsilon)\rho(x)} \quad on \quad \mathbf{R}^n \tag{2}$$

for any $\varepsilon > 0$ where $\rho(x)$ is some function which tends to infinity as $|x| \to \infty$ and which reflects the different behavior of the potential V in different directions The study of estimates such as (2) for solutions of second order elliptic equations in unbounded domains will be our main task in these lectures An important technical point in such a study is a good choice of the function ρ in (2) which should describe as closely as possible the decaying pattern of a whole class of solutions of the equation $Pu = \mu u$ in some neighborhood of infinity

What makes eigenfunctions with eigenvalues below the bottom of the essential spectrum decay exponentially? The answer to this question is to be found in the observation that exponential decay properties of eigenfunctions and other solutions of the equation $Pu = \mu u$ are closely related to positivity properties of the quadratic form $((P-\mu)\varphi,\varphi)$ on certain subsets of test functions To see this connection we invoke a formula for the bottom of the essential spectrum of H given by Persson [32] (see also Chapter 3) by which

$$\Sigma = \sup_K \inf\{(P\varphi,\varphi) \quad \varphi \in C_0^\infty(\mathbf{R}^n \setminus K) \, , \, \|\varphi\| = 1\} \tag{3}$$

where the supremum is taken over the family of compact subsets $K \subset \mathbf{R}^n$ It follows from (3) that if $\mu < \Sigma$ and ε is any positive number such that $\varepsilon < \Sigma - \mu$ then there exists a number $R = R_\varepsilon > 0$ such that

$$((P-\mu)\varphi,\varphi) \geq (\Sigma-\mu-\varepsilon) \int_{|x|>R} |\varphi|^2 dx \tag{4}$$

for all $\varphi \in C_0^\infty(\Omega_R)$ where $\Omega_R = \{x \in \mathbf{R}^n \quad |x| > R\}$ Thus we see in particular that for $\mu < \Sigma$ the quadratic form $((P-\mu)\varphi,\varphi)$ is strongly positive over all $\varphi \in C_0^\infty(\Omega_R)$ for R sufficiently large We now claim that the quadratic form lower bound (4) implies the exponential decay upper bound (1) for eigenfunctions and other solutions of the equation $Pu = \mu u$. This follows from results to be established in these lectures.

The principle which states that positivity of the quadratic form $((P-\mu)\varphi,\varphi)$ on $C_0^\infty(\Omega_R)$ implies decay of solutions of $Pu = \mu u$ can be extended to yield non-isotropic exponential decay estimates. Thus suppose that (4) is replaced by a more general inequality of the form

$$((P-\mu)\varphi,\varphi) \geq \int_{|x|>R} \lambda(x)|\varphi|^2\, dx \tag{5}$$

for all $\varphi \in C_0^\infty(\Omega_R)$ where $\lambda(x)$ is some positive continuous function on $\mathbb{R}^n$. Suppose (for simplicity) that $0 < c_1 \leq \lambda(x) \leq c_2(1+|x|)^N$ for some constants c_1, c_2 and N. Let $\rho(x)$ denote the geodesic distance from x to the origin in the Riemannian metric

$$ds^2 = \lambda(x)(dx_1^2 + \cdots + dx_n^2). \tag{6}$$

It will be shown in these lectures that if $\psi(x)$ is a solution of $P\psi = \mu\psi$ in Ω_R which does not increase too fast in the sense that

$$\int_{\Omega_R} |\psi|^2 e^{-2(1-\delta)\rho(x)} dx < \infty$$

for some $\delta > 0$, then in fact ψ decays exponentially in the following L^2 sense:

$$\int_{\Omega_R} |\psi|^2 e^{2(1-\varepsilon)\rho(x)} dx < \infty \tag{7}$$

for any $\varepsilon > 0$. Under a mild additional restriction on V_- we shall moreover deduce from (7) the non-isotropic pointwise estimate (2) for x in Ω_R.

The last result is applicable with some modifications to N-body Schrödinger operators. It yields non-isotropic exponential decay upper bounds for eigenfunctions of various multiparticle quantum systems. The bounds are of the form (2) with $\rho(x)$ defined as the geodesic distance in a Riemannian metric similar to (6) with a certain multiplier $\lambda(x)$ defined by the system (we note that λ depends only on $x/|x|$ and that it is in general a discontinuous function of direction). The non-isotropic exponential decay upper bounds thus obtained are precise and cannot be improved in an essential way in case the eigenfunction ψ is the *ground state* (i.e. when μ is the lowest eigenvalue) of an N-body Schrodinger operator This follows from a recent work by Carmona and Simon [6] on ***lower bounds*** for such eigenfunctions

These lectures are oriented toward applications to Schrodinger differential equations on $\mathbb{R}^n$ The method used can however be adapted to yield exponential decay results in other situations which are of interest Thus one can obtain results on exponential decay of solutions of elliptic equations in an unbounded domain Ω with a ***non-compact*** boundary $\partial\Omega$ assuming that the solutions vanish on $\partial\Omega$ outside some compact Generalizing in another direction one can study by the same method the rate of decay of solutions of elliptic equations on non-compact Riemannian manifolds

The plan of these notes is as follows In Chapter 0 we introduce various function spaces and recall some related technical results In Chapter 1 we present the basic weighted L^2 estimates for solutions of second order elliptic equations $Pu = f$ in a domain Ω in $\mathbb{R}^n$ These estimates are proved in Theorem 1 5 under a λ-positivity assumption on the form $\mathrm{Re}(P\varphi,\varphi)$ and a growth restriction on u The theorem is preceded by a discussion of some elementary properties of Riemannian (and Finsler) metric functions which appear as ***weights*** in the main estimates Chapter 2 prepares the ground for applications of Theorem 1 5 The discussion in this chapter is motivated by the following question related to the "λ-Condition" Given an elliptic operator P on $\mathbb{R}^n$ find a function $\lambda(x)$ which is as large as possible such that for some neighborhood of infinity Ω_R

$$(P\varphi,\varphi) \geq \int_{\Omega_R} \lambda(x)|\varphi|^2 dx$$

for all $\varphi \in C_0^\infty(\Omega_R)$ Lemma 2 8 gives an answer to this question under the additional requirement that λ is as a homogeneous function of degree 0 Chapter 3 deals with the self-adjointness problem for elliptic operators on $\mathbb{R}^n$ The basic results on the existence of self-adjoint realizations of such operators and their spectral properties are given in Theorem 3 2 Chapter 4 contains the main examples of L^2 exponential decay results of these lectures Typical examples are the non-isotropic exponential decay estimates for

for eigenfunctions of atomic type Schrödinger operators given in Theorem 4.12 and the non-isotropic exponential decay estimates for eigenfunctions of N-body Schrödinger operators given in Theorem 4.13. Chapter 5 deals with pointwise estimates. The aim in this chapter is to show that the various L^2 exponential decay results established in Chapter 4 imply similar ***pointwise*** estimates (under somewhat stronger assumptions). Examples of such pointwise estimates for eigenfunctions of multiparticle Schrödinger operators are given in Theorem 5.2 and Theorem 5.3. The key result in Chapter 5 is Theorem 5.1 which gives pointwise bounds for weak solutions of second order elliptic equations with measurable coefficients and possibly complex valued lower order terms. The estimates are extensions of similar estimates established by Stampacchia which in turn generalize the well known pointwise bounds for solutions of second order elliptic equations in divergence form given by De Giorgi and Nash. The notes conclude with four appendices and with a section of bibliographical comments.

We remark that some of the results on exponential decay of eigenfunctions of multiparticle Schrödinger operators as well as other results given in these lectures were announced by us in [1] and in [2]. The present volume gives complete details of these results together with some additional material.

I am greatly indebted to Emanuela Caliceti, Richard Froese, Ira Herbst and James Howland for their help in preparing these lectures for publication. Not only did they take the notes of these lectures, in itself no easy task, but they have also improved the original presentation by supplementing the notes in several places with some of the missing details. It is with a great pleasure that I acknowledge this help and express my thanks for the work done.

Last and not least special thanks are due to Danit Sharon for her very skillful computer editing on a DEC VAX 11/780 under the UNIX (a trademark of Bell Laboratories) "troff" package.

Shmuel Agmon

Chapter 0 Preliminaries

In this chapter we introduce various function spaces which play a role in the sequel and state some of their basic properties. Let Ω be an open connected subset of $\mathbb{R}^n$. $B(x;r)$ will denote an open ball in $\mathbb{R}^n$ centered at $x = (x_1, \ldots, x_n)$ and with radius r. Let ∂_i denote $\frac{\partial}{\partial x_i}$.

If $F(\Omega)$ is a space of functions on Ω, then $F_0(\Omega)$ denotes members of $F(\Omega)$ with compact support.

If for every open subset U of Ω we are given a space $F(U)$ of functions on U, then $F_{loc}(\Omega)$ denotes functions f such that for every $x \in \Omega$ there exists a neighborhood U of x such that $f \restriction U \in F(U)$.

$C^k(\Omega)$ denotes the space of k times continuously differentiable functions. The members of $C_0^\infty(\Omega)$ (space of infinitely differentiable functions with compact support) are sometimes called test functions.

$L^p(\Omega)$ denotes the space of (equivalence classes of) functions which are measurable and have integrable p-th power with respect to the Lebesgue measure on Ω.

$L^\infty(\Omega)$ denotes the space of essentially bounded functions.

$H^1(\Omega)$ denotes the Sobolev space of functions $u \in L^2(\Omega)$ such that the distributional derivatives $\partial_i u$ are in $L^2(\Omega)$ for $i = 1, \ldots, n$.

For $0 \leq \delta < 1$ we introduce the space of functions $M_\delta(\Omega)$ defined as follows. For $n \geq 3$ (recall that n is the dimension of the underlying space) or for $n = 2$ and $\delta \neq 0$, the space $M_\delta(\Omega)$ consists of the functions $u \in L^1_{loc}(\Omega)$ such that

$$\lim_{r\to 0} \int_{B(x^0;r)\cap\Omega} |u(x)|\, |x-x^0|^{2-n-\delta} dx = 0$$

uniformly in $x^0 \in \mathbb{R}^n$. In the case $n = 2$, $\delta = 0$, replace $|x-x^0|^{2-n-\delta}$ with $|\log |x-x^0||$ in the defining expression, and when $n = 1$ replace $|x-x^0|^{2-n-\delta}$ with 1.

$M_\delta(\Omega)$ is an example of a Stummel space of functions [41]. For most of the results given in these lectures we need to consider only the spaces $M_0(\Omega)$ and $M_{0,loc}(\Omega)$ (the space $M_\delta(\Omega)$ with $\delta > 0$ will be needed however in the last chapter). In the following we shall simplify notation and write $M(\Omega)$ for the space $M_0(\Omega)$ and $M_{loc}(\Omega)$ for the space $M_{0,loc}(\Omega)$. Note that by Hölder's inequality $M_{loc}(\Omega) \supset L^p_{loc}(\Omega)$ for $p > n/2$ and $n \geq 2$, and for $n = 1$ $M_{loc}(\Omega) = L^1_{loc}(\Omega)$.

Finally a real valued function on Ω is Lipschitz if for every $x^0 \in \Omega$ there exists a neighborhood U of x^0 and a number $C > 0$ such that

$$|u(x)-u(y)| \leq C|x-y| \qquad \text{for all} \quad x,y \in U.$$

We will have occasion to use the following two results:

0.1 Rademacher's Theorem: *If u is a Lipschitz function then the differential of u exists almost everywhere.*

This theorem is proved in [27; Theorem 3.1.6, p. 65].

In the following $\Lambda^s\varphi$ is defined for $\varphi \in C_0^\infty(\mathbb{R}^n)$ (or also for $\varphi \in \mathcal{S}(\mathbb{R}^n)$) via the Fourier transform map $\mathcal{F} : L^2(\mathbb{R}^n) \to L^2(\mathbb{R}^n)$, with $\mathcal{F}u = \hat{u}$, by

$$\mathcal{F}(\Lambda^s\varphi)(\xi) = (1+|\xi|^2)^{s/2}\hat{\varphi}(\xi) \qquad \text{on} \quad \mathbb{R}^n.$$

(Thus in particular $\Lambda = (1-\Delta)^{1/2}$ where $\Delta = \sum_{i=1}^{n} \partial_i^2$ is the Laplacian.)

Also, in the following lemmas $\|\cdot\|$ denotes the norm in $L^2(\mathbb{R}^n)$ and functions defined in $\Omega \subset \mathbb{R}^n$ are assumed to have been extended as zero in $\mathbb{R}^n\setminus\Omega$.

0.2 Lemma: *Let $g \in M_{\delta,loc}(\Omega)$. Then for every $\varepsilon > 0$ and every compact subset K of Ω there exists a constant $C(\varepsilon,K)$ such that for $\vartheta = 1-(\delta/2)$*

$$|| \, |g|^{1/2}\varphi|| \leq \varepsilon||\Lambda^{\vartheta}\varphi|| + C(\varepsilon,K)||\varphi|| \tag{0.1}$$

for all $\varphi \in C_0^\infty(\Omega)$ with $\operatorname{supp}\varphi \subset K$. When $g \in M_\delta(\Omega)$ the constant $C(\varepsilon,K)$ can be chosen independently of K.

The proof of this Lemma can be found in [36, Theorem 7 3, p. 138]. When $\delta = 0$ we obtain the following corollary

0.3 Lemma: *Suppose $g \in M_{loc}(\Omega)$ Then for any $\varepsilon > 0$ and every compact subset K of Ω there exist constants $C_1(\varepsilon,K)$ and $C_2(\varepsilon,K)$ such that*

$$|| \, |g|^{1/2}\varphi|| \leq \varepsilon||\nabla\varphi|| + C_1(\varepsilon,K)||\varphi||, \tag{0.2}$$

$$\int_\Omega |g|\,|\varphi|^2dx \leq \varepsilon\int_\Omega \sum_{i=1}^n |\partial_i\varphi|^2dx + C_2(\varepsilon,K)\int_\Omega |\varphi|^2dx \tag{0.3}$$

for all $\varphi \in C_0^\infty(\Omega)$ with $\operatorname{supp}\varphi \subset K$ Here $||\nabla u||$ denotes $(\int_\Omega \sum_{i=1}^n |\partial_i u|^2dx)^{1/2}$. If $g \in M(\Omega)$ C_1 and C_2 can be chosen independently of K.

Chapter 1 The Main Theorem

Let Ω be a connected open set in $\mathbb{R}^n$. We consider in Ω a second order elliptic operator $A(x,\partial)$ of the form

$$A(x,\partial) = -\sum_{i,j=1}^{n} \partial_j a^{ij}(x)\partial_i \tag{1.1}$$

where $a^{ij}(x)$ are continuous real valued functions and the matrix $[a^{ij}(x)]$ is positive definite for every $x \in \Omega$. We shall consider the equation

$$A(x,\partial)u(x) + q(x)u(x) = f(x) \tag{1.2}$$

for complex valued functions $q \in L^1_{loc}(\Omega)$ and $f \in L^2_{loc}(\Omega)$. Since the coefficients $a^{ij}(x)$ are not required to be differentiable the equation (1.2) should be interpreted in a generalized sense. In this connection it is convenient to introduce the following notation:

1.1 Definition: *Let $u,v \in H^1_{loc}(\Omega)$. Then we define*

$$\nabla_A u(x) \cdot \nabla_A v(x) = \sum_{i,j=1}^{n} a^{ij}(x)\partial_i u(x)\partial_j v(x), \tag{1.3}$$

$$|\nabla_A u(x)|^2 = \nabla_A u(x) \cdot \nabla_A \overline{u(x)} = \sum_{i,j=1}^{n} a^{ij}(x)\partial_i u(x)\partial_j \overline{u(x)}. \tag{1.4}$$

With this notation a function $u \in H^1_{loc}(\Omega)$ is said to be a solution of (1.2) if $qu \in L^1_{loc}(\Omega)$ and

$$\int_\Omega (\nabla_A u \cdot \nabla_A \varphi + qu\varphi)dx = \int_\Omega f\varphi dx \tag{1.5}$$

for every $\varphi \in C_0^\infty(\Omega)$. Note that the integral on the left side of (1.5) is obtained by formally integrating by parts the expression $\int_\Omega \varphi Au dx$. Note also that for any $u \in H^1_{loc}(\Omega)$ $A(x,\partial)u$ is defined as the distribution T on $C_0^\infty(\Omega)$ such that

$$T(\varphi) = \int_\Omega \nabla_A u \cdot \nabla_A \varphi \, dx \tag{1.6}$$

for all $\varphi \in C_0^\infty(\Omega)$.

The goal of this chapter is to prove Theorem 1 5 which is the main technical tool in these notes This theorem states that under certain conditions, solutions of equation (1 2) which do not grow too fast in fact decay rapidly The main condition which is required for the validity of the theorem is a positivity condition on the quadratic form associated with the operator $A+q$ More precisely, it is required that there exist a positive continuous function λ on Ω such that

$$\mathrm{Re} \int_\Omega (|\nabla_A \varphi|^2 + q(x)|\varphi|^2)dx \geq \int_\Omega \lambda(x)|\varphi|^2 dx \tag{1 7}$$

for all $\varphi \in C_0^\infty(\Omega)$ We shall call this condition the λ-Condition The function λ appearing in this condition will be used to define a Riemannian distance function ρ_λ on Ω which will be used to measure decay (or growth) of solutions of (1 2) As a preparation to the main theorem we shall first define these metric functions and prove some of their properties (We note that these properties will be proved in Lemma 1 3 for the more general class of *Finsler metrics* The more general results may turn out to be useful in other situations)

Let A be the elliptic operator on Ω defined by (1 1) and let $\lambda(x)$ be a continuous positive function on Ω With the pair $\{A, \lambda\}$ we associate a Riemannian metric ds_λ^2 on Ω defined by

$$ds_\lambda^2 = \lambda(x) \sum_{i,j=1}^{n} a_{ij}(x)dx_i dx_j \tag{1 8}$$

where $[a_{ij}(x)]$ is the inverse of the matrix $[a^{ij}(x)]$ The distance between points x and y in Ω in this metric is given by

$$\rho_\lambda(x,y) = \inf_\gamma \int_0^1 [\lambda(\gamma(t))\sum_{i,j} a_{ij}(\gamma(t))\gamma_i(t)\gamma_j(t)]^{1/2} dt \tag{1 9}$$

where the infimum is taken over all absolutely continuous paths γ $[0,1] \to \Omega$ such that $\gamma(0) = y$ and $\gamma(1) = x$ For $E \subset \Omega$ let $\rho_\lambda(x,E) = \inf\{\rho_\lambda(x,y)$ $y \in E\}$ Then if $\Omega \cup \{\infty\}$ is the one point compactification of Ω we define for $x \in \Omega$

$$\rho_\lambda(x,\{\infty\}) = \sup_K \{\rho_\lambda(x,\Omega/K) \quad K \text{ is a compact subset of } \Omega\} \tag{1 9)'$$

In Appendix 1 it is shown that the metric space (Ω, ρ_λ) is complete if and only if $\rho_\lambda(x,\{\infty\}) = \infty$ for some and hence, by the triangle inequality, for every $x \in \Omega$. The point $\{\infty\}$ is called the ideal boundary of Ω. It is easy to see that if ρ_λ is the Euclidean metric, $\rho_\lambda(x,\{\infty\})$ is indeed the distance from x to the boundary of Ω.

The following Lemmas develop some properties of metrics where the function $[\lambda(x)\sum_{i,j} a_{ij}(x)\xi_i\xi_j]^{1/2}$ appearing in the definition of ρ_λ is replaced by a more general function $K_*(x,\xi)$. These metrics are known as Finsler metrics [11].

1.2 Lemma: *Let $K(x,\xi)$ be a function on $\Omega \times \mathbb{R}^n$ satisfying*

(1) *K is continuous in x and ξ.*

(2) *$K > 0$ for $\xi \neq 0$.*

(3) *$K(x,a\xi) = aK(x,\xi)$ for $a > 0$, i.e., K is positively homogeneous of degree 1.*

(4) *K is convex in ξ.*

Define K_*, the polar function of K, by

$$K_*(x,\xi) = \sup_{\eta \neq 0} \frac{<\xi,\eta>}{K(x,\eta)}$$

where $<\cdot,\cdot>$ denotes the usual inner product in $\mathbb{R}^n$ given by $<\xi,\eta> = \sum_{i=1}^{n} \xi_i\eta_i$.

Then K_* also enjoys properties (1) to (4). Also $K_{**} = K$.

This Lemma is proved in Appendix 2.

Given $K_*(x,\xi)$ the associated Finsler metric on Ω is defined by

$$\rho_{K_*}(x,y) = \inf_\gamma \int_0^1 K_*(\gamma(t),\dot\gamma(t))dt \tag{1.10}$$

where the infimum is taken over all absolutely continuous paths such that $\gamma(0) = y$, $\gamma(1) = x$. Although we call it a metric, the Finsler metric is not necessarily symmetric, since it is not true in general that $K_*(x,\xi) = K_*(x,-\xi)$. Nevertheless, the triangle in equality holds in the following form

$$\rho_{K_*}(x,z) \le \rho_{K_*}(x,y) + \rho_{K_*}(y,z)$$

as is easily shown by restricting the set of paths from z to x to those which also pass through y.

1.3 Lemma: *Let K satisfy the hypotheses of Lemma 1.2 and let K_* be as defined there. Let $\rho_{K_*}(x,y)$ be the associated Finsler metric given by (1.10). Then*

(i) *For any fixed y $\rho_{K_*}(x,y)$ is Lipschitz in x and*

$$K(x,\nabla\rho_{K_*}(x,y)) \le 1 \text{ a.e.}$$

(ii) *If h is any real Lipschitz function then*

$K(x,\nabla h(x)) \le 1$ a.e. *iff $h(x)-h(y) \le \rho_{K_*}(x,y)$ for all $x,y \in \Omega$.*

Proof: (i) Choose a ball $B(x^0;r) \subset \overline{B}(x^0;r) \subset \Omega$. Then for $x^1,x^2 \in B(x^0;r)$

$$\begin{aligned}\rho_{K_*}(x^1,x^0) - \rho_{K_*}(x^2,x^0) &\le \rho_{K_*}(x^1,x^2)\\ &= \inf_\gamma \int_0^1 K_*(\gamma(t),\dot\gamma(t))dt\\ &\le \int_0^1 K_*(tx^1+(1-t)x^2,x^1-x^2)dt\\ &= \int_0^1 |x^1-x^2| K_*(tx^1+(1-t)x^2 , \frac{x^1-x^2}{|x^1-x^2|})dt\\ &\le |x^1-x^2| \sup\{K_*(x,\xi) : (x,\xi) \in \overline{B}(x^0;r) \times \overline{B}(0;1)\}.\end{aligned}$$

The first inequality is the triangle inequality for ρ_{K_*}. Continuity of K_* guarantees the existence of the supremum over a compact set in the final expression. This and a similar argument with x^1 and x^2 interchanged proves that $\rho_{K_*}(x,y)$ is Lipschitz in x.

Now let $x^0 \in \Omega$ and define

$$\rho_{K_*}(x) = \rho_{K_*}(x,x^0).$$

Since ρ_{K_*} is Lipschitz it has a differential almost everywhere by Rademacher's theorem. Let x be a point where $\nabla\rho_{K_*}$ exists. By the triangle inequality

$$\rho_{K_*}(x+s\omega)-\rho_{K_*}(x) \leq \rho_{K_*}(x+s\omega,\, x) \leq \int_0^1 K_*(x+st\omega,\, s\omega)dt$$

where $\omega \in \mathbb{R}^n$ and $s > 0$ is sufficiently small so that $x+t\omega \in \Omega$ if $0 \leq t \leq s$.

Since $K_*(x,\xi)$ is positive homogeneous of degree 1 in ξ

$$\frac{\rho_{K_*}(x+s\omega)-\rho_{K_*}(x)}{s} \leq \int_0^1 K_*(x+st\omega,\, \omega)dt$$

so taking the limit as $s \to 0$

$$\frac{<\nabla\rho_{K_*}(x),\omega>}{K_*(x,\omega)} \leq 1.$$

Finally

$$K(x,\nabla\rho_{K_*}(x)) = K_{**}(x,\nabla\rho_{K_*}(x)) = \sup_{\omega\neq 0}\frac{<\nabla\rho_{K_*}(x),\omega>}{K_*(x,\omega)} \leq 1$$

which proves (i).

(ii) Suppose $K(x,\nabla h(x)) \leq 1$ a.e. We wish to show $h(x)-h(y) \leq \rho_{K_*}(x,y)$. To begin we assume $h \in C^1(\Omega)$ so that $K(x,\nabla h(x)) \leq 1$ for all $x \in \Omega$. Then if γ is an absolutely continuous path from y to x in Ω, $h(\gamma(t))$ is absolutely continuous and

$$\frac{d}{dt}h(\gamma(t)) = <\nabla h(\gamma(t)),\dot\gamma(t)>$$

as an L^1 function. The inequality

$$<\xi,\eta> \leq K(x,\xi)K_*(x,\eta) \quad \text{for} \quad x \in \Omega \;\wedge\; \xi,\eta \in \mathbb{R}^n$$

follows easily from the definition of K_*. Therefore

$$h(x)-h(y) = \int_0^1 \frac{d}{dt} h(\gamma(t))dt = \int_0^1 <\nabla h(\gamma(t)),\dot\gamma(t)> dt$$

$$\leq \int_0^1 K(\gamma(t),\nabla h(\gamma(t)))\ K_*(\gamma(t),\dot\gamma(t))dt$$

$$\leq \int_0^1 K_*(\gamma(t),\dot\gamma(t))dt.$$

The last inequality follows because by hypothesis $K(\gamma(t),\nabla h(\gamma(t))) \leq 1$. Taking the infimum over all absolutely continuous paths γ we obtain

$$h(x)-h(y) \leq \rho_{K_*}(x,y).$$

Now consider the general case when h is Lipschitz but not necessarily C^1. Suppose $x^0,y^0 \in \Omega$. Given $\varepsilon > 0$ suppose γ is an absolutely continuous path from x^0 to y^0 with

$$\int_0^1 K_*(\gamma(t),\dot\gamma(t))dt \leq (1+\varepsilon)\rho_{K_*}(x,y).$$

Since γ is continuous, $\Gamma = \{\gamma(t) : t \in [0,1]\}$ is compact. Therefore we can find an open subset Ω' of Ω such that $\Gamma \subset \Omega'$ and for some $\varepsilon_1 > 0$ $B(x;\varepsilon_1) \subset \Omega$ for all $x \in \Omega'$. Let ζ be a non negative test function on $\mathbb{R}^n$ such that $\zeta(x) = 0$ if $|x| > 1$ and

$$\int_{\mathbb{R}^n} \zeta(x)dx = 1.$$

Define for $0 < \varepsilon < \varepsilon_1$

$$\zeta_\varepsilon = \varepsilon^{-n}\ \zeta(\frac{x}{\varepsilon}). \tag{1.11}$$

Then the following smoothed version of h is defined for $x \in \Omega'$:

$$h_\varepsilon(x) = \int_{\mathbb{R}^n} \zeta_\varepsilon(y)h(x-y)dy.$$

As $\varepsilon \to 0$, $h_\varepsilon \to h$ uniformly on compact subsets of Ω'.

Now we must estimate $K(x,\nabla h_\varepsilon(x))$ for $x \in \Gamma$ under the assumption $K(x,\nabla h(x)) \leq 1$.

$$\frac{<\nabla h_\varepsilon(x),\eta>}{K_*(x,\eta)} = \int \zeta_\varepsilon(y)\frac{<\nabla h(x-y),\eta>}{K_*(x,\eta)}dy$$
$$\leq \int \zeta_\varepsilon(y)\frac{K_*(x-y,\eta)K(x-y,\nabla h(x-y))}{K_*(x,\eta)}dy$$
$$\leq \int \zeta_\varepsilon(y)\frac{K_*(x-y,\eta)}{K_*(x,\eta)}dy.$$

Since K_* is continuous and bounded away from zero if $|\eta| = 1$, we can take the supremum over $|\eta| = 1$ in the previous inequality. Noting that supp $\zeta_\varepsilon \subset B(0;\varepsilon)$ we find that the right side of the above inequality converges to 1 uniformly as $\varepsilon \to 0$. So for $x \in \Gamma$

$$\sup_{|\eta|=1}\frac{<\nabla h_\varepsilon(x),\eta>}{K_*(x,\eta)} \leq 1 + \delta(\varepsilon)$$

where $\delta(\varepsilon) \to 0$ as $\varepsilon \to 0$. The homogeneity of K_* implies that

$$\sup_{|\eta|=1}\frac{<\nabla h_\varepsilon(x),\eta>}{K_*(x,\eta)} = \sup_{\eta\neq 0}\frac{<\nabla h_\varepsilon(x),\eta>}{K_*(x,\eta)} = K(x,\nabla h_\varepsilon(x)) \leq 1 + \delta(\varepsilon).$$

Now h_ε is certainly differentiable and since $\Gamma \subset \Omega'$, $h_\varepsilon(\gamma(t))$ is well defined. Therefore we can mimic the original argument to obtain

$$h_\varepsilon(x^0)-h_\varepsilon(y^0) \leq (1+\delta(\varepsilon))(1+\varepsilon)\rho_{K_*}(x^0,y^0)$$

so as $\varepsilon \to 0$ we reach the desired conclusion.

Conversely let h be a real Lipschitz function satisfying for all $x,y \in \Omega$: $h(x)-h(y) \leq \rho_{K_*}(x,y)$. Let x be a point where ∇h exists, then

$$h(x+s\omega)-h(x) \leq \rho_{K_*}(x+s\omega,\, x) \leq \int_0^1 K_*(x+st\omega,s\omega)dt$$

where s is positive and sufficiently small and $\omega \in \mathbb{R}^n\setminus\{0\}$.

Now the same argument used to prove (i) applies with ρ_{K_*} replaced by h. Since K_* is positively homogeneous of degree 1

$$\frac{h(x+s\omega)-h(x)}{s} \leq \int_0^1 K_*(x+st\omega,\, \omega)dt$$

and taking the limit as $s \to 0$

$$\frac{<\nabla h(x),\omega>}{K_*(x,\omega)} \le 1$$

Finally, $K(x,\nabla h(x)) = K_{**}(x,\nabla h(x)) = \sup_{\omega\neq 0} \frac{<\nabla h(x),\omega>}{K_*(x,\omega)} \le 1$ and (ii) is proved ▪

1.4 Theorem: *Suppose $a^{ij}(x)$, $1 \le i, j \le n$, are continuous real valued functions such that $[a^{ij}(x)]$ is a positive definite matrix for every x. Let $\lambda(x)$ be a positive continuous function. Let $|\nabla_A u|^2$ and $\rho_\lambda(x,y)$ be as defined by (1 4) and (1 9). Then*

(i) $|\nabla_A \rho_\lambda(x)|^2 \le \lambda(x)$ *where* $\rho_\lambda(x) = \rho_\lambda(x,y)$ *for some fixed* $y \in \Omega$

(ii) *If h is a real Lipschitz function then* $|\nabla_A h(x)|^2 \le \lambda(x)$ a e *iff* $h(x)-h(y) \le \rho_\lambda(x,y)$ *for all* $x,y \in \Omega$

Remark: The inequality (ii) is called the eikonal inequality. For additional discussion of this and the associated metric see [8, Chapter II, §9]

Proof. Let $K(x,\xi) = [\lambda(x)^{-1}\sum_{i,j} a^{ij}(x)\xi_i\xi_j]^{1/2}$. The verification of properties (1) through (4) of section 1 2 is immediate. We only remark that convexity follows from the fact that for each fixed x, $K(x,\ )$ is a norm. In fact if $B(x) = \lambda(x)^{-1/2}[a^{ij}(x)]^{1/2}$ we have $K(x,\xi) = ||B(x)\xi|| = <B(x)\xi,B(x)\xi>^{1/2}$. The corresponding polar function K_* is given by

$$\begin{aligned} K_*(x,\xi) &= \sup_{\eta\neq 0}(<\xi,\eta>/\, K(x,\eta)) \\ &= \sup_{\eta\neq 0}(<\xi,\eta>/\, ||B(x)\eta||) \\ &= \sup_{\eta\neq 0}(<B(x)^{-1}\xi,B(x)\eta>/\, ||B(x)\eta||) \\ &= ||B(x)^{-1}\xi|| \\ &= <\xi,B(x)^{-2}\xi>^{1/2} \\ &= [\lambda(x)\sum_{i,j} a_{ij}(x)\xi_i\xi_j]^{1/2}, \end{aligned}$$

where the fourth equality is a consequence of Schwarz's inequality. Therefore $\rho_\lambda(x,y) = \rho_{K_*}(x,y)$. Since $K(x,\nabla h(x))^2 = \lambda(x)^{-1}|\nabla_A h(x)|^2$, (i) and (ii) are direct consequences of Lemma 1 3 ▪

We pass to the main theorem of this chapter.

1.5 Theorem: *Let* Ω *be a connected open set in* $\mathbb{R}^n$ *and let* $A(x,\partial) = -\sum_{i,j}\partial_j a^{ij}(x)\partial_i$ *be the elliptic operator on* Ω *introduced in* (1.1). *Let* $q(x)$ *be a complex valued function on* Ω *such that*

(i) $q \in L^1_{loc}(\Omega)$

(ii) $q_- \in M_{loc}(\Omega)$ *where* $q_-(x) = \max(0, -\mathrm{Re}\, q(x))$.

Suppose that there exists a positive continuous function $\lambda(x)$ *on* Ω *such that*

$$\mathrm{Re}\int_\Omega (|\nabla_A\varphi|^2 + q(x)|\varphi|^2)dx \geq \int_\Omega \lambda(x)|\varphi|^2 dx \tag{1.12}$$

for every $\varphi \in C_0^\infty(\Omega)$. *Let* $\rho_\lambda(x,y)$ *be the geodesic distance in* Ω *between the points* x, y *in the Riemannian metric*

$$ds_\lambda^2 = \lambda(x) \sum_{i,j=1}^n a_{ij}(x)dx_i dx_j,$$

$[a_{ij}] = [a^{ij}]^{-1}$. *Fix a point* $y^0 \in \Omega$ *and set* $\rho_\lambda(x) = \rho_\lambda(x,y^0)$. *Let* h *be a real valued Lipschitz function on* Ω *such that*

$$|\nabla_A h(x)|^2 < \lambda(x) \text{ a.e.}$$

Suppose now that u *is a function in* $H^1_{loc}(\Omega)$ *which satisfies the differential equation*

$$Au + qu = f$$

in the sense that $qu \in L^1_{loc}(\Omega)$, $f \in L^2_{loc}(\Omega)$, *and* (1.5) *holds. Suppose also that*

$$\int_\Omega |u(x)|^2\lambda(x)e^{-2(1-\delta)\rho_\lambda(x)}dx < \infty \tag{1.13}$$

for some $\delta > 0$.

Then the following inequalities hold:

(a) *If* Ω *is complete in the metric* ρ_λ, *then*

$$\int_\Omega |u(x)|^2(\lambda(x) - |\nabla_A h(x)|^2)e^{2h(x)}dx \tag{1.14}$$

$$\le \int_\Omega |f(x)|^2(\lambda(x)-|\nabla_A h(x)|^2)^{-1}e^{2h(x)}dx$$

(b) *In general, if Ω is not necessarily complete, let $d > 0$ and define $\Omega_d = \{x \in \Omega \ \ \rho_\lambda(x,\{\infty\}) > d\}$ where $\rho_\lambda(x,\{\infty\})$ is defined by* (1 9) *Then*

$$\int_{\Omega_d} |u(x)|^2(\lambda(x)-|\nabla_A h(x)|^2)e^{2h(x)}dx$$
$$\le \int_\Omega |f(x)|^2(\lambda(x)-|\nabla_A h(x)|^2)^{-1}e^{2h(x)}dx \qquad (1\ 15)$$
$$+ \frac{2(1+2d)}{d^2}\int_{\Omega\setminus\Omega_d} |u(x)|^2\lambda(x)e^{2h(x)}dx$$

Remark Suppose that $\mathbb{R}^n\setminus\Omega$ is a compact set and that ρ_λ has the following properties

(i) For a fixed $y^0 \in \Omega$ $\rho_\lambda(x,y^0) \to \infty$ as $|x| \to \infty$, $x \in \Omega$

(ii) $\rho_\lambda(x,y)$ admits a continuous extension from $\Omega \times \Omega$ to $\overline{\Omega} \times \overline{\Omega}$

In this situation it is easy to see that $\rho_\lambda(x,\{\infty\})$ coincides with the distance from x to the boundary of Ω, so that in this case

$$\Omega_d = \{x \in \Omega \ , \ \rho_\lambda(x,\partial\Omega) > d\}$$

Since the proof of this theorem is long and the trend of the argument is somewhat obscured by the details we first give a short discussion of the ideas involved

We begin by deriving an identity to be used below Let $v \in C^1(\Omega)$, $\psi \in C^1(\Omega)$, ψ real Then a calculation shows

$$\nabla_A v \ \nabla_A(\psi^2\bar{v}) = |\nabla_A(\psi v)|^2 - |v|^2|\nabla_A\psi|^2 - [\psi v\nabla_A\psi \ \nabla_A\bar{v} - \psi\bar{v}\nabla_A v \ \nabla_A\psi]$$

Since the last term is pure imaginary,

$$\mathrm{Re}[\nabla_A v \ \nabla_A(\psi^2\bar{v})] = |\nabla_A(\psi v)|^2 - |v|^2|\nabla_A\psi|^2 \qquad (1\ 16)$$

Now suppose u is a solution to the differential equation Then we have

$$\int_\Omega \nabla_A u \ \ \nabla_A\varphi dx + \int_\Omega qu\varphi dx = \int_\Omega f\varphi dx$$

for every $\varphi \in C_0^\infty(\Omega)$. Suppose we could replace φ with $\bar{u}\psi^2$ where ψ has compact support. Then we would obtain, after taking the real part and using identity (1.16)

$$\int_\Omega [|\nabla_A(\psi u)|^2 - |u|^2|\nabla_A\psi|^2 + \mathrm{Re}q\,|u|^2\psi^2]dx \tag{1.16)'$$

$$= \mathrm{Re}\int_\Omega f\bar{u}\psi^2 dx.$$

Invoking the λ-condition (1.12) with $\varphi = \psi u$ we would get

$$\int_\Omega [\lambda|u\psi|^2 - |u|^2|\nabla_A\psi|^2]dx \le \mathrm{Re}\int_\Omega f\bar{u}\psi^2 dx.$$

In fact as we shall later show this inequality is valid if ψ is a real Lipschitz function of compact support. It appears as (1.20) below.

Suppose we ignore the requirement that ψ have compact support and use $\psi = e^h$ in the above inequality. Then proceeding formally we obtain

$$\int_\Omega |u|^2(\lambda - |\nabla_A h|^2)e^{2h}dx \le \mathrm{Re}\int_\Omega f\bar{u}e^{2h}dx.$$

An application of Schwarz's inequality to the right side

$$\mathrm{Re}\int_\Omega f\bar{u}e^{2h}dx = \mathrm{Re}\int_\Omega f\bar{u}(\lambda - |\nabla_A h|^2)^{1/2}(\lambda - |\nabla_A h|^2)^{-1/2}e^{2h}dx$$

$$\le \{\int_\Omega |u|^2(\lambda - |\nabla_A h|^2)e^{2h}dx\}^{1/2}\{\int_\Omega |f|^2(\lambda - |\nabla_A h|^2)^{-1}e^{2h}dx\}^{1/2}$$

produces, after dividing both sides of the resulting inequality by $\{\int_\Omega |u|^2(\lambda - |\nabla_A h|^2)e^{2h}\}^{1/2}$ and squaring both sides,

$$\int_\Omega |u|^2(\lambda - |\nabla_A h|^2)e^{2h}dx$$

$$\le \int_\Omega |f|^2(\lambda - |\nabla_A h|^2)^{-1}e^{2h}dx$$

which is precisely the result wanted when (Ω, ρ_λ) is complete.

Note that we did not use the growth assumption on u with the illegal choice $\psi = e^h$. If we now try to approximate e^h with legal choices for ψ, for example $\psi = e^h\chi$ with χ a smoothed characteristic function of a compact set (which will eventually become large), then quantities involving the derivatives of χ must be estimated in the remote regions of Ω where the growth of u will be important This is precisely what happens in the proof of the theorem, to which we now turn

Proof By hypothesis u is a solution of $A(x,\partial)u + qu = f$, i e, $u \in H^1_{loc}(\Omega)$, $qu \in L^1_{loc}(\Omega)$ and

$$\int_\Omega \nabla_A u \cdot \nabla_A \varphi dx + \int_\Omega qu\varphi dx = \int_\Omega f \varphi dx \qquad (1\ 17)$$

for $\varphi \in C_0^\infty(\Omega)$ By a density argument this equation is still true if $\varphi \in H^1(\Omega) \cap L_0^\infty(\Omega)$ For suppose $\varphi \in H^1(\Omega) \cap L_0^\infty(\Omega)$ If ζ_ε is as in the proof of Lemma 1 3 (see (1 11)) then for $0 < \varepsilon \le \varepsilon_1$ $\varphi_\varepsilon(x) = \int_\Omega \zeta_\varepsilon(x-y)\varphi(y)dy$ is in $C_0^\infty(\Omega)$ with support in a fixed compact set In addition $\varphi_\varepsilon \to \varphi$ in $H^1(\Omega)$ as $\varepsilon \to 0$ while $||\varphi_\varepsilon||_{L^\infty} \le ||\varphi||_{L^\infty}$ Using φ_ε in (1 17) we see that the first and last terms converge as $\varepsilon \downarrow 0$ while going to a sequence $\{\varepsilon_i\}$, $\varepsilon_i \downarrow 0$, for which $\varphi_{\varepsilon_i} \to \varphi$ a e, the second term converges by Lebesgue's dominated convergence theorem

Let $u_\varepsilon = u/(1+\varepsilon|u|^2)$ The distributional derivatives of u_ε can be calculated as if u were differentiable in the ordinary sense This is easily seen by first approximating u by a smooth function and going back to the definition of distributional derivative A short calculation also shows that $u_\varepsilon \to u$ in $H^1_{loc}(\Omega)$ as $\varepsilon \downarrow 0$

If ψ is a real Lipschitz function of compact support an additional approximation argument can be used to show that $\bar{u}_\varepsilon\psi^2 \in H^1(\Omega) \cap L_0^\infty(\Omega)$ and that $\partial_i(\bar{u}_\varepsilon\psi^2) = (2\psi\partial_i\psi)\bar{u}_\varepsilon + \psi^2\partial_i\bar{u}_\varepsilon$ We thus set $\varphi = \bar{u}_\varepsilon\psi^2$ and substitute into (1 17) The result is

$$\int_\Omega \nabla_A u \cdot \nabla_A(\bar{u}_\varepsilon\psi^2)dx + \int_\Omega qu\bar{u}_\varepsilon\psi^2 dx = \int_\Omega f\bar{u}_\varepsilon\psi^2 dx$$

Next we replace u with $u_\varepsilon + (u - u_\varepsilon)$ and take the real part Then the equation assumes the form

$$\mathrm{Re}\int_\Omega \nabla_A u_\varepsilon \cdot \nabla_A(\bar{u}_\varepsilon \psi^2)dx + \mathrm{Re}\int_\Omega q\,|u_\varepsilon|^2\psi^2 dx = \mathrm{Re}\int_\Omega f\bar{u}_\varepsilon\psi^2 dx + I_\varepsilon \tag{1.18}$$

where

$$I_\varepsilon = \mathrm{Re}\int_\Omega \nabla_A(u_\varepsilon - u)\cdot\nabla_A(\bar{u}_\varepsilon\psi^2)dx + \mathrm{Re}\int_\Omega q\,(u_\varepsilon - u)\bar{u}_\varepsilon\psi^2 dx$$

$$= \mathrm{Re}\int_\Omega \nabla_A(u_\varepsilon - u)\cdot\nabla_A(\bar{u}_\varepsilon\psi^2)dx + \int_\Omega (q_- - q_+)(u - u_\varepsilon)\bar{u}_\varepsilon\psi^2 dx.$$

(Here $\mathrm{Re}\, q = q_+ - q_-$ with $q_+ = \max(0, \mathrm{Re}\, q)$.) We now estimate I_ε. We first note that $(q_- - q_+)(u - u_\varepsilon)\bar{u}_\varepsilon \le q_-(u - u_\varepsilon)\bar{u}_\varepsilon$ and then use Schwarz's inequality to obtain

$$I_\varepsilon \le \{\int_{supp\,\psi} |\nabla_A(u - u_\varepsilon)|^2 dx\}^{1/2}\{\int_{supp\,\psi} |\nabla_A(\psi^2 u_\varepsilon)|^2 dx\}^{1/2}$$

$$+ \{\int_{supp\,\psi} q_-\psi^2|u - u_\varepsilon|^2 dx\}^{1/2}\{\int_{supp\,\psi} \psi^2 q_-|u_\varepsilon|^2 dx\}^{1/2}.$$

Since $u_\varepsilon \to u$ in H^1_{loc} as $\varepsilon \to 0$ and $supp\ \psi$ is compact the first factor of the first term tends to zero as $\varepsilon \to 0$. The second factor stays bounded for the same reasons. To estimate the factors of the second term we recall that $q_- \in M_{loc}(\Omega)$ and invoke lemma 0.3. This gives for any fixed $\delta > 0$

$$\int_{supp\,\psi} q_-\psi^2|u - u_\varepsilon|^2 dx \le \delta \int_{supp\,\psi} \sum_i |\partial_i(\psi(u - u_\varepsilon))|^2 dx$$

$$+ C(\delta, supp\,\psi)\int_{supp\,\psi} \psi^2|u - u_\varepsilon|^2 dx,$$

which implies that the first factor of the second term tends to zero as $\varepsilon \to 0$ while a similar estimate shows that the second factor remains bounded. Therefore

$$\limsup_{\varepsilon\to 0}\ I_\varepsilon \le 0.$$

Now the identity (1.16) is used to rewrite equation (1.18) as follows:

$$\int_\Omega [|\nabla_A(\psi u_\varepsilon)|^2 + \mathrm{Re}\, q\,|u_\varepsilon|^2\psi^2 - |u_\varepsilon|^2|\nabla_A\psi|^2]dx \tag{1.19}$$

$$= \mathrm{Re}\int_\Omega f\bar{u}_\varepsilon\psi^2 dx + I_\varepsilon.$$

The condition (1.12) on $\lambda(x)$ implies, by a density argument

$$\int_\Omega [|\nabla_A(\psi u_\varepsilon)|^2 + \mathrm{Re}\, q\, |u_\varepsilon|^2\psi^2]dx \geq \int_\Omega \lambda |\psi u_\varepsilon|^2 dx.$$

This inequality is applied to equation (1.19):

$$\int_\Omega [\lambda |\psi u_\varepsilon|^2 - |u_\varepsilon|^2 |\nabla_A \psi|^2]dx \leq \mathrm{Re}\int_\Omega f \bar{u}_\varepsilon \psi^2 dx + I_\varepsilon.$$

Let $\varepsilon \to 0$ to obtain the important inequality

$$\int_\Omega [\lambda |u\psi|^2 - |u|^2 |\nabla_A \psi|^2]dx \leq \mathrm{Re}\int_\Omega f \bar{u} \psi^2 dx. \tag{1.20}$$

We now use our freedom in choosing ψ. Recall from the remark preceding the proof we want ψ to approximate $e^{h(x)}$. With this in mind let $\psi(x) = e^{g(x)}\chi(x)$ where g and χ are real Lipschitz functions, χ has compact support and $0 \leq \chi \leq 1$. Suppose in addition that $|\nabla_A g|^2 < \lambda$ a.e. Then

$$|\nabla_A \psi|^2 = e^{2g}\chi^2 |\nabla_A g|^2 + 2e^{2g}\chi\, \nabla_A \chi \cdot \nabla_A g + e^{2g} |\nabla_A \chi|^2.$$

So (1.20) implies

$$\int_\Omega |u\chi|^2(\lambda - |\nabla_A g|^2)e^{2g} \leq \mathrm{Re}\int_\Omega f\bar{u}x^2 e^{2g} dx + \int_\Omega |u|^2 e^{2g}(|\nabla_A \chi|^2 + 2\chi |\nabla_A \chi \cdot \nabla_A g|)dx.$$

The first term on the right is estimated as follows:

$$\mathrm{Re}\int_\Omega f\bar{u}\chi^2 e^{2g} dx = \mathrm{Re}\int_\Omega f\bar{u}\chi^2 e^{2g}(\lambda - |\nabla_A g|^2)^{1/2}(\lambda - |\nabla_A g|^2)^{-1/2} dx$$

$$\leq \{\int_\Omega |u\chi|^2(\lambda - |\nabla_A g|^2)e^{2g} dx\}^{1/2}\{\int_\Omega |f\chi|^2(\lambda - |\nabla_A g|^2)^{-1} e^{2g} dx\}^{1/2}.$$

Therefore

$$\int_\Omega |u\chi|^2(\lambda - |\nabla_A g|^2)e^{2g} dx \leq$$

$$\{\int_\Omega |u\chi|^2(\lambda - |\nabla_A g|^2)e^{2g} dx\}^{1/2}\{\int_\Omega |f\chi|^2(\lambda - |\nabla_A g|^2)]^{-1} e^{2g} dx\}^{1/2}$$

$$+ \int_\Omega |u|^2 e^{2g}(|\nabla_A \chi|^2 + 2\chi |\nabla_A \chi \cdot \nabla_A g|)dx.$$

This inequality is of the form

$$a \leq a^{1/2} b^{1/2} + c$$

where a b and c are positive. We claim this implies

$$a \leq b + 2c.$$

Certainly if $c \geq a$ this is true On the other hand if $a > c$ then

$$a - c \leq a^{1/2} b^{1/2} \text{ implies } a^2 - 2ac + c^2 \leq ab$$

so

$$a + \frac{c^2}{a} \leq b + 2c$$

which implies the result Therefore

$$\int_\Omega |u\chi|^2(\lambda - |\nabla_A g|^2)e^{2g}\,dx \leq \int_\Omega |f\chi|^2(\lambda - |\nabla_A g|^2)^{-1}e^{2g}\,dx$$
$$+ 2\int_\Omega |u|^2 e^{2g}(|\nabla_A \chi|^2 + 2\chi|\nabla_A \chi\, \nabla_A g|)\,dx \qquad (1\ 21)$$

(Note that we do not exclude the trivial case when the first integral on the right side of (1 21) is infinite)

We now let functions of the form $e^{g(x)}\chi(x)$ approximate $e^{h(x)}$ This will be done by constructing sequences $\chi_{d,j}$ and h_i which satisfy the hypotheses for χ and g respectively, inserting these functions in (1 21) and taking limits At the same time we will try to control the last term in (1 21)

Let $\{K_i\}$ be a sequence of compact sets in Ω such that $K_i \subset K_j$ if $i < j$ and $\bigcup_{i=1}^{\infty}$ (interior K_i) $= \Omega$ Given $d > 0$, define the function $\eta_d(t)$ on $[0,\infty)$ as follows

$$\eta_d(t) = \begin{cases} t/d & \text{if } t \in [0,d] \\ 1 & \text{if } t \in (d,\infty), \end{cases}$$

and let

$$\chi_{d,j}(x) = \eta_d(\rho_\lambda(x, \Omega \setminus K_j))$$

Note that outside K_j $\chi_{d,j} = 0$ and inside K_j $\chi_{d,j} = 1$ except for points which are closer than d in the metric ρ_λ to the exterior of K_j On these points $\chi_{d,j} = \rho_\lambda(x, \Omega \setminus K_j)/d$ Thus $\chi_{d,j}$ has compact support and $0 \leq \chi_{d,j} \leq 1$ Furthermore $\chi_{d,j}$ is Lipschitz To see this note that for $t, t_1 \in [0,\infty)$

$$|\eta_d(t) - \eta_d(t_1)| \leq d^{-1}|t - t_1|$$

Thus we have

$$|\chi_{d,j}(x) - \chi_{d,j}(y)| \le d^{-1}|\rho_\lambda(x, \Omega\setminus K_j) - \rho_\lambda(y, \Omega\setminus K_j)|$$

$$\le d^{-1}\rho_\lambda(x,y) \tag{1 22}$$

$$\le d^{-1}C|x-y|$$

where given x^0 the last inequality holds for x and y in some neighborhood of x^0 and for some constant C This follows from the definitions of ρ_λ The second inequality is a consequence of the triangle inequality So $\chi_{d,j}$ can be used in (1 21)

To control the final term in (1 21) we need to estimate the derivatives of $\chi_{d,j}$ To this end, note that by the second inequality of (1 22) $d\chi_{d,j}$ satisfies the assumptions of (ii) in Theorem 1 4 Therefore

$$|\nabla_A \chi_{d,j}(x)|^2 \le d^{-2}\lambda(x) \tag{1 23}$$

We now turn to the construction of the h_i Recall there is $\delta > 0$ such that

$$\int_\Omega |u(x)|^2\lambda(x)e^{-2(1-\delta)\rho_\lambda(x)}dx < \infty$$

With no loss of generality we assume that $0 < \delta < 1$ We use this δ to define the sequence of functions

$$h_i(x) = \min\{h(x), -(1-\delta)\rho_\lambda(x) + i\}, \quad i = 1,2,$$

Clearly h_i is Lipschitz We now estimate derivatives of h_i Since for any i all three functions $h(x)$, $-(1-\delta)\rho_\lambda(x) + i$ and $h_i(x)$ are Lipschitz, for almost every x the differentials of all three functions exist At such a point x^0 the differential of h_i is equal to either the differential of h or that of $-(1-\delta)\rho_\lambda(x) + i$ This is clearly true if $-(1-\delta)\rho_\lambda(x^0) + i \ne h(x^0)$ On the other hand if $-(1-\delta)\rho_\lambda(x^0) + i = h(x^0)$ then for $\omega \in \mathbb{R}^n$ and $t > 0$ small enough

$$\frac{h_i(x^0+t\omega)-h_i(x^0)}{t} \le \frac{h(x^0+t\omega)-h(x^0)}{t}$$

so that sending t to zero gives

$$<\omega, [\nabla h_i(x^0) - \nabla h(x^0)]> \le 0$$

for all $\omega \in \mathbb{R}^n$ This implies $\nabla h_i(x^0) = \nabla h(x^0)$ Therefore at a point where the differentials of all three functions exists either

$$|\nabla_A h_i|^2 = |\nabla_A h|^2$$

or

$$|\nabla_A h_i|^2 = (1-\delta)^2 |\nabla_A \rho_\lambda|^2$$

Now $|\nabla_A h|^2 < \lambda$ a e by hypothesis and by Theorem 1 4
$(1-\delta)^2 |\nabla_A \rho_\lambda|^2 \le (1-\delta)^2 \lambda < \lambda$ a e Thus we find

$$|\nabla_A h_i|^2 < \lambda \quad a\,e \tag{1 24}$$

Therefore h_i is suitable for use in (1 21)

The estimates (1 24) and (1 23) can be combined to yield

$$|\nabla_A \chi_{d,j}\ \nabla_A h_i| \le |\nabla_A \chi_{d,j}|\ |\nabla_A h_i| \le \frac{1}{d}\lambda \tag{1 25}$$

where $|\nabla_A f|$ denotes $(|\nabla_A f|^2)^{1/2}$

We return to inequality (1 21) inserting $\chi = \chi_{d,j}$ and $g = h_i$

$$\begin{aligned} &\int_\Omega |u \chi_{d,j}|^2 (\lambda - |\nabla_A h_i|^2) e^{2h_i} dx \\ &\le \int_\Omega |f \chi_{d,j}|^2 (\lambda - |\nabla_A h_i|^2)^{-1} e^{2h_i} dx \\ &+ 2\int_\Omega e^{2h_i} |u|^2 (|\nabla_A \chi_{d,j}|^2 + 2\chi_{d,j} |\nabla_A \chi_{d,j}\ \nabla_A h_i|) dx \end{aligned} \tag{1 26}$$

Note that the last term on the right involves derivatives of $\chi_{d,j}$ which vanish except on $K_j \setminus K_{j,d}$, where $K_{j,d} = \{x \in K_j \ \ \rho_\lambda(x, \Omega \setminus K_j) > d\}$ Let $\mathcal{K}_{j,d}$ be the characteristic function of $K_j \setminus K_{j,d}$

$$\mathcal{K}_{j,d}(x) = \begin{cases} 1 & \text{if } x \in K_j \setminus K_{j,d} \\ 0 & \text{if } x \notin K_j \setminus K_{j,d} \end{cases}$$

Recall that $\Omega_d = \{x \in \Omega \ \ \rho_\lambda(x, \{\infty\}) > d\}$ Since any compact set K is contained in interior K_j for large enough j,

$$\lim_{j \to \infty} \mathcal{K}_{j,d}(x) = 0 \quad \text{if} \quad x \in \Omega_d$$

Now we rewrite the last term of (1 26)

$$2\int_{\Omega} e^{2h_\imath}|u|^2(|\nabla_A\chi_{d,j}|^2 + 2\chi_{d,j}|\nabla_A\chi_{d,j}\,\nabla_A h_\imath|)dx$$

$$= 2\int_{\Omega\setminus\Omega_d} e^{2h_\imath}|u|^2(|\nabla_A\chi_{d,j}|^2 + 2\chi_{d,j}|\nabla_A\chi_{d,j}\,\nabla_A h_\imath|)dx$$

$$+ 2\int_{\Omega_d}\mathcal{K}_{j,d}(x)e^{2h_\imath}|u|^2(|\nabla_A\chi_{d,j}|^2 + 2\chi_{d,j}|\nabla_A\chi_{d,j}\,\nabla_A h_\imath|)dx$$

$$+ 2\int_{\Omega_d}(1-\mathcal{K}_{j,d}(x))e^{2h_\imath}|u|^2(|\nabla_A\chi_{d,j}|^2 + 2\chi_{d,j}|\nabla_A\chi_{d,j}\,\nabla_A h_\imath|)dx$$

The integrand of the third integral on the right vanishes identically It follows from the definition of $h_\imath$ and the growth assumption (1 13) that

$$\int_{\Omega}|u(x)|^2\lambda(x)\,e^{2h_\imath(x)}dx < \infty$$

Therefore the estimates (1 23) and (1 25) allow us to apply Lebesgue's dominated convergence theorem to the second integral on the right which thus tends to zero as $j \to \infty$ Therefore, letting $j \to \infty$ in (1 26) we find that

$$\limsup_{j\to\infty} \int_{\Omega}|u\chi_{d,j}|^2(\lambda-|\nabla_A h_\imath|^2)e^{2h_\imath}dx$$

$$\leq \limsup_{j\to\infty} \int_{\Omega}|f\chi_{d,j}|^2(\lambda-|\nabla_A h_\imath|^2)^{-1}e^{2h_\imath}dx \tag{1 27}$$

$$+ 2\limsup_{j\to\infty} \int_{\Omega\setminus\Omega_d} e^{2h_\imath}|u|^2(|\nabla_A\chi_{d,j}|^2 + 2\chi_{d,j}|\nabla_A\chi_{d,j}\,\nabla_A h_\imath|)dx$$

$$\leq \int_{\Omega}|f|^2(\lambda-|\nabla_A h_\imath|^2)^{-1}e^{2h_\imath}dx$$

$$+ \frac{2(1+2d)}{d^2}\int_{\Omega\setminus\Omega_d} e^{2h}|u|^2\lambda\,dx$$

where we have used $0 \leq \chi_{d,j} \leq 1$, the estimates (1 23) and (1 25), and $h_\imath \leq h$ to derive the second inequality

Since $\chi_{d,j} \to 1$ on Ω_d an application of Fatou's lemma gives

$$\int_{\Omega_d}|u|^2(\lambda-|\nabla_A h_\imath|^2)e^{2h_\imath}dx$$

$$\leq \limsup_{j\to\infty}\int_{\Omega_d}|u\chi_{d,j}|^2(\lambda-|\nabla_A h_\imath|^2)e^{2h_\imath}dx$$

Thus combining the last inequality with (1 27), we get

$$\int_{\Omega_d}|u|^2(\lambda-|\nabla_A h_\imath|^2)e^{2h_\imath}dx$$

$$\leq \int_\Omega |f|^2(\lambda-|\nabla_A h_i|^2)^{-1}e^{2h_i}dx \qquad (1\ 28)$$

$$+\frac{2(1+2d)}{d^2}\int_{\Omega\setminus\Omega_d} e^{2h}|u|^2\lambda\, dx$$

We now claim that $(\lambda-|\nabla_A h_i|^2)^{-1} \leq \delta^{-1}(\lambda-|\nabla_A h|^2)^{-1}$ This is immediate at points where $|\nabla_A h_i|^2 = |\nabla_A h|^2$ At points where $|\nabla_A h_i|^2 = (1-\delta)^2|\nabla_A \rho_\lambda|^2$

$$\begin{aligned}\lambda-|\nabla_A h_i|^2 &= \lambda-(1-\delta)^2|\nabla_A\rho_\lambda|^2\\ &\geq \lambda-(1-\delta)^2\lambda\\ &= \delta\lambda(2-\delta) \geq \delta\lambda\\ &\geq \delta(\lambda-|\nabla_A h|^2)\end{aligned}$$

Thus if $\delta^{-1}\int_\Omega |f|^2(\lambda-|\nabla_A h|^2)^{-1}e^{2h}dx < \infty$ we can apply Lebesgue's dominated convergence theorem to the first term on the right of (1 28) This gives

$$\lim_{i\to\infty}\int_\Omega |f|^2(\lambda-|\nabla_A h_i|^2)^{-1}e^{2h_i}dx = \int_\Omega |f|^2(\lambda-|\nabla_A h|^2)^{-1}e^{2h}dx \qquad (1\ 29)$$

Finally letting $i\to\infty$ in (1 28), applying Fatou's lemma on the left side of (1 28) and using (1 29), we get

$$\begin{aligned}&\int_{\Omega_d} |u|^2(\lambda-|\nabla_A h|^2)e^{2h}dx\\ &\leq \limsup_{i\to\infty}\int_{\Omega_d}|u|^2(\lambda-|\nabla_A h_i|^2)e^{2h_i}dx\\ &\leq \limsup_{i\to\infty}\int_\Omega |f|^2(\lambda-|\nabla_A h_i|^2)^{-1}e^{2h_i}dx + \frac{2(1+2d)}{d^2}\int_{\Omega\setminus\Omega_d}|u|^2\lambda e^{2h}dx\\ &\leq \int_\Omega |f|^2(\lambda-|\nabla_A h|^2)^{-1}e^{2h}dx + \frac{2(1+2d)}{d^2}\int_{\Omega\setminus\Omega_d}|u|^2\lambda e^{2h}dx\end{aligned}$$

Here we have assumed that the first integral on the right side of the final inequality is finite to apply the previous remark Otherwise the inequality is trivially true

This completes the proof of part (b) of the theorem Part (a) is simply a special case because when Ω is complete in the metric ρ_λ, $\Omega = \Omega_d$ ▪

We conclude this chapter with a simple example where Theorem 1 5 is applied Consider the Schrodinger equation (associated with the Stark effect):

$$-\Delta u + (V(x)+x_1)u = 0 \tag{1.30}$$

where Δ is the Laplacian on $\mathbb{R}^n$, $V \in L^1_{loc}(\mathbb{R}^n)$, $V_- \in M_{loc}(\mathbb{R}^n)$ and Re $V \to 0$ as $x_1 \to +\infty$. Suppose that u is a solution of (1.30) in the half-space: $\mathbb{R}^n_+ = \{x : x \in \mathbb{R}^n , x_1 > 0\}$ (as usual we assume that $u \in H^1_{loc}(\mathbb{R}^n_+)$ and $Vu \in L^1_{loc}(\mathbb{R}^n_+)$). If $u \in L^2(\mathbb{R}^n_+)$, then

$$\int_{\mathbb{R}^n_+} |u(x)|^2 e^{\frac{4}{3}(x_1-\varepsilon)_+^{3/2}} dx < \infty \tag{1.31}$$

for every $\varepsilon > 0$.

To prove this fix $\varepsilon > 0$ and choose $N > \varepsilon$ such that $|\mathrm{Re}V(x)| < \varepsilon/2$ for $x_1 > N$. Set $\Omega = \{x : x_1 > N\}$. For $\varphi \in C_0^\infty(\Omega)$ we have:

$$\int_\Omega (|\nabla\varphi|^2 + \mathrm{Re}(V+x_1)|\varphi|^2)dx \geq \int_\Omega (x_1-\frac{\varepsilon}{2})|\varphi|^2 dx.$$

Thus in Ω the operator $-\Delta+V+x_1$ satisfies the conditions of Theorem 1.5 with $\lambda(x) = x_1-\varepsilon/2 > 0$. Let $\rho_\lambda(x,y)$ be the distance in Ω between the points $x = (x_1, \ldots, x_n)$ and $y = (y_1, \ldots, y_n)$ in the Riemannian metric:

$$ds^2 = (x_1-\frac{\varepsilon}{2})(dx_1^2 + \cdots + dx_n^2).$$

Fix y and set $\rho_\lambda(x) = \rho_\lambda(x,y)$. Since

$$\rho_\lambda(x) \geq |\int_{y_1}^{x_1} (t-\frac{\varepsilon}{2})^{\frac{1}{2}} dt|,$$

it is clear that $\lambda(x)e^{-2(1-\delta)\rho_\lambda(x)}$ is a bounded function for any $\delta \in (0,1)$. Thus it follows that u verifies the growth condition (1.13) in Theorem 1.5.

We can choose $h(x) = \frac{2}{3}(x_1-\varepsilon)^{3/2}$ and note that

$$|\nabla h(x)|^2 = |\frac{2}{3}\partial_1(x_1-\varepsilon)^{3/2}|^2 = \lambda(x) - \frac{\varepsilon}{2}, \tag{1.32}$$

so that h verifies the condition: $|\nabla h(x)|^2 < \lambda(x)$ a.e. which was required in Theorem 1.5. Applying the theorem, using (1.15) with $d = 1$, we obtain

$$\int_{\Omega_1} |u(x)|^2(\lambda(x) - |\nabla h(x)|^2)e^{2h(x)}dx \leq 6 \int_{\Omega \setminus \Omega_1} |u(x)|^2\lambda(x)e^{2h(x)}dx \tag{1.33}$$

where $\Omega_1 = \{x : x \in \Omega \, , \rho_\lambda(x, \partial\Omega) > 1\}$. Now it is easily seen that $\Omega_1 = \{x : x_1 > N_1\}$ for some $N_1 > N$ which implies that the function $\lambda(x)\ \exp(2h(x))$, which depends only on x_1, is bounded in $\Omega \setminus \Omega_1$. Thus, using (1.32) it follows from (1.33) that

$$\int_{\Omega_1} |u(x)|^2 e^{\frac{4}{3}(x_1 - \varepsilon)^{3/2}} dx \leq Const. \int_{\Omega \setminus \Omega_1} |u|^2 dx < \infty$$

which implies (1.31).

Chapter 2 Geometric Spectral Analysis

Our goal is to apply Theorem 1.5 to measure the decay of solutions to $Au+qu=0$. If $q(x)=q_1(x)-\mu$ we thereby obtain estimates for eigenfunctions of $A+q_1$ with eigenvalue μ. To apply Theorem 1.5 the real part of the quadratic form associated with the operator $A+q$ need be positive on $C_0^\infty(\Omega)$. Moreover, we must find a positive continuous function $\lambda(x)$ such that the quadratic form in question is strongly positive on $C_0^\infty(\Omega)$ in the sense that the λ-Condition (1.7) holds. (Since we are dealing with behavior of solutions "at infinity" the inequality (1.7) need to hold only for test functions supported in $\Omega\setminus K$ for some compact set K.) In this chapter we introduce various quantities related to the spectrum of operators obtained by restricting $A+q$ to various subsets of Ω. As we shall see these quantities will allow us under certain conditions to produce the desired λ functions.

To simplify the discussion we assume from now on, unless otherwise specified, that $A(x,\partial)$ is the elliptic operator (1.1) defined on $\Omega=\mathbb{R}^n$. We shall assume that the coefficients $a^{ij}(x)$ of A are continuous *bounded* functions on $\mathbb{R}^n$. Again let $q\in L^1_{loc}(\mathbb{R}^n)$ and $q_-\in M_{loc}(\mathbb{R}^n)$ and assume that q is real valued. With these assumptions set

$$P(x,\partial)=A(x,\partial)+q(x) \tag{2.1}$$

and define

$$(Pu,u)=\int_{\mathbb{R}^n}(|\nabla_A u|^2+q\,|u|^2)dx \tag{2.2}$$

for all $u\in H^1_{loc}(\mathbb{R}^n)$ for which the integral (2.2) makes sense. Our main aim in this chapter is to find continuous functions $\lambda(x)$ on $\mathbb{R}^n$, not necessarily positive but "as large as possible", such that the following inequality holds

$$(P\varphi,\varphi)\geq\int_{\mathbb{R}^n}\lambda|\varphi|^2dx \tag{2.3}$$

for all $\varphi\in C_0^\infty(\mathbb{R}^n)$.

From now on unless otherwise specified, the norm $\|\cdot\|$ denotes the norm in $L^2(\mathbb{R}^n)$.

2.1 Definition: *Let P be the operator introduced above. For any $y \in \mathbb{R}^n$ and $R > 0$ define*

$$\Lambda_R(y;P) = \inf\{\frac{(P\varphi,\varphi)}{\|\varphi\|^2} : \varphi \in C_0^\infty(B(y;R)), \varphi \neq 0\} \tag{2.4}$$

where $B(y;R)$ is the ball of radius R centered at y.

Note that $\Lambda_R(y;P)$ can be identified with the lowest eigenvalue of the self-adjoint realization of P in $L^2(B(y;R))$ under zero Dirichlet boundary conditions. This property will not be used in the sequel. On the other hand we shall use the following properties of Λ_R.

2.2 Lemma: *$\Lambda_R(x;P)$ is a continuous function of (x,R) on $\mathbb{R}^n \times \mathbb{R}_+$. Furthermore $\Lambda_R(x;P) = \Lambda_R(x;A+q)$ is also continuous in $[a^{ij}]$ in the sense that if $A_m = -\sum_{i,j=1}^{n} \partial_j a_m^{ij} \partial_i$ where $[a_m^{ij}(x)]$ has all the properties of $[a^{ij}(x)]$ for $m = 1,2,\cdots$, and $\lim_{m\to\infty} |a^{ij}(x) - a_m^{ij}(x)| = 0$ uniformly on compact sets, then $\Lambda_R(x;A_m+q) \to \Lambda_R(x;A+q)$ uniformly for x in compact subsets.*

A proof of Lemma 2.2 is given in Appendix 3.

We now produce functions $\lambda(x)$ which verify (2.3).

2.3 Lemma: *Under the above conditions on P for any $\varepsilon > 0$ there exists $R_\varepsilon > 0$ such that*

$$(P\varphi,\varphi) \geq \int_{\mathbb{R}^n} (\Lambda_R(x;P) - \varepsilon)|\varphi(x)|^2 dx \tag{2.5}$$

for all $\varphi \in C_0^\infty(\mathbb{R}^n)$ and for any $R \geq R_\varepsilon$.

Proof: Let ζ be real valued function in $C_0^\infty(\mathbb{R}^n)$ such that

$$\zeta(x) = 0 \quad \text{if} \quad |x| > \frac{1}{2}$$

$$\int_{\mathbb{R}^n} \zeta^2 = 1.$$

For $R > 0$ and $y \in \mathbb{R}^n$ let

$$\zeta_{R,y}(x) = \zeta(\frac{x-y}{R}),$$

$$\zeta_R(x) = \zeta_{R,0}(x)$$

Since the derivatives of ζ are bounded in $\mathbb{R}^n$, we have

$$|\nabla_A \zeta_{R,y}(x)|^2 \le \frac{C}{R^2} \tag{2 6}$$

for some constant $C > 0$ depending only on ζ and the upper bound for the a^{ij} Temporarily assume that $a^{ij} \in C^1(\mathbb{R}^n)$ Then $f = P\varphi$ is well defined and in $L_0^1(\mathbb{R}^n)$ if $\varphi \in C_0^\infty(\mathbb{R}^n)$

Clearly

$$\int_{\mathbb{R}^n} \nabla_A \varphi \quad \nabla_A(\zeta_{R,y}^2 \bar\varphi)dx + \int_{\mathbb{R}^n} q\,|\varphi|^2 \zeta_{R,y}^2 dx$$
$$= \int_{\mathbb{R}^n} f\, \zeta_{R,y}^2 \bar\varphi dx$$

Using identity (1 16) this leads to

$$\int_{\mathbb{R}^n} (|\nabla_A(\zeta_{R,y}\varphi)|^2 + q\,|\zeta_{R,y}\varphi|^2)dx - \int_{\mathbb{R}^n} |\nabla_A \zeta_{R\,y}|^2 |\varphi|^2 dx \tag{2 7}$$
$$= \mathrm{Re} \int_{\mathbb{R}^n} f\, \zeta_{R,y}^2 \bar\varphi dx$$

The first term is recognized and estimated as follows

$$\int_{\mathbb{R}^n} (|\nabla_A(\zeta_{R\,y}\varphi)|^2 + q\,|\zeta_{R,y}\varphi|^2)dx$$
$$= (P(\zeta_{R,y}\varphi), \zeta_{R,y}\varphi) \tag{2 8}$$
$$\ge \Lambda_{R/2}(y,P) \int_{B(y\;R/2)} |\varphi\zeta_{R,y}|^2 dx$$

Now suppose $|x-y| \le R/2$ Clearly, $B(y,R/2) \subset B(x,R/2+|x-y|) \subset B(x,R)$ Therefore

$$\Lambda_{R/2}(y,P) \ge \Lambda_{R/2+|x-y|}(x,P) \ge \Lambda_R(x,P) \tag{2 9}$$

The equations (2 6) through (2 9) together with the fact that $\zeta_R(x-y)$ is supported in $B(y,R/2)$ imply

$$\int_{\mathbb{R}^n} \Lambda_R(x,P)|\varphi(x)|^2 \zeta_R^2(x-y)dx - CR^{-2} \int_{B(y,R)} |\varphi(x)|^2 dx$$

$$\leq \operatorname{Re} \int_{\mathbb{R}^n} f(x)\overline{\varphi}(x)\, \zeta_R^2(x-y)dx.$$

Integration with respect to y gives

$$R^n \int_{\mathbb{R}^n} \Lambda_R(x;P)|\varphi(x)|^2dx - CR^{-2}V_1R^n \int_{\mathbb{R}^n} |\varphi(x)|^2dx$$
$$\leq R^n \operatorname{Re} \int_{\mathbb{R}^n} f(x)\overline{\varphi}(x)dx$$

where V_1 is the volume of the ball of radius 1 in $\mathbb{R}^n$. Dividing by R^n and recalling the definition of f we get

$$(P\varphi,\varphi) \geq \int_{\mathbb{R}^n} \Lambda_R(x;P)|\varphi(x)|^2dx - CR^{-2}V_1 \int_{\mathbb{R}^n} |\varphi(x)|^2dx \tag{2.10}$$

and the lemma for the case $a^{ij} \in C^1$ follows by taking R_ε such that $R_\varepsilon^2 = CV_1\varepsilon^{-1}$.

Since C only depends on the upper bound for the a^{ij} (and on ζ), we can approximate $[a^{ij}(x)]$ in the case when the a^{ij} are not necessarily C^1 uniformly by a sequence of positive definite matrices $[a_m^{ij}]$ with smooth entries which are uniformly bounded from above (see Lemma A1.1, Appendix 1). Then by Lemma 2.2 $\Lambda_R(x;A_m+q) \to \Lambda_R(x;P)$ uniformly on compact sets so that

$$\int_{\mathbb{R}^n} \Lambda_R(x;A_m+q)|\varphi(x)|^2dx \to \int_{\mathbb{R}^n} \Lambda_R(x;P)|\varphi(x)|^2dx \quad as \quad m \to \infty.$$

Also $((A_m+q)\varphi,\varphi) \to (P\varphi,\varphi)$ as $m \to \infty$. Since the a_m^{ij} are uniformly bounded from above we can choose the constant C such that for every m the inequality (2.10) holds with $P = A_m+q$. Thus (2.10) is also valid for $P = A+q$ and the lemma follows. ▪

The following quantities will play an important role in what follows.

2.4 Definition:

$$\Lambda(P) = \inf\left\{\frac{(P\varphi,\varphi)}{\|\varphi\|^2} : \varphi \in C_0^\infty(\mathbb{R}^n),\ \varphi \neq 0\right\}, \tag{2.11}$$

$$\Sigma(P) = \sup_K \inf\left\{\frac{(P\varphi,\varphi)}{\|\varphi\|^2} : \varphi \in C_0^\infty(\mathbb{R}^n \setminus K)\ ,\ \varphi \neq 0\right\} \tag{2.12}$$

where the supremum is taken over the family of compact subsets $K \subset \mathbb{R}^n$.

$\Lambda(P)$ and $\Sigma(P)$ will be shown to be equal to the bottom of the spectrum and essential spectrum respectively of the self-adjoint realization of P on $L^2(\mathbb{R}^n)$ when $\Sigma(P) > -\infty$. In general both may take on the value $-\infty$.

It is easy to see that $\Lambda(P) = \lim_{R\to\infty} \Lambda_R(x;P)$ for any $x \in \mathbb{R}^n$. Note that since $\Lambda_R(x;P)$ is a decreasing function of R this limit exists (although it might be $-\infty$). The relationship between $\Sigma(P)$ and $\Lambda_R(x;P)$ is the subject of the following lemma.

2.5 Lemma:

$$\Sigma(P) = \lim_{R\to\infty} \liminf_{|x|\to\infty} \Lambda_R(x;P).$$

Proof: Let K be a compact subset of $\mathbb{R}^n$ and $R > 0$. Then for x such that $|x|$ is sufficiently large $B(x;R) \subset \mathbb{R}^n \setminus K$. Therefore for such an x,

$$\inf\left\{\frac{(P\varphi,\varphi)}{\|\varphi\|^2} : \varphi \in C_0^\infty(\mathbb{R}^n \setminus K)\ ,\ \varphi \neq 0\right\} \le \Lambda_R(x;P)$$

and hence

$$\inf\left\{\frac{(P\varphi,\varphi)}{\|\varphi\|^2} : \varphi \in C_0^\infty(\mathbb{R}^n \setminus K)\ ,\ \varphi \neq 0\right\} \le \liminf_{|x|\to\infty} \Lambda_R(x;P). \tag{2.13}$$

Since the right hand side of (2.13) does not depend on K and the left hand side does not depend on R it follows from (2.12) that

$$\Sigma(P) \le \lim_{R\to\infty} \liminf_{|x|\to\infty} \Lambda_R(x;P). \tag{2.14}$$

By Lemma 2.3, given $\varepsilon > 0$ there exists $R_\varepsilon > 0$ such that

$$(P\varphi,\varphi) \ge \int_{\mathbb{R}^n} (\Lambda_R(x;P) - \varepsilon)\,|\varphi(x)|^2 dx \tag{2.15}$$

for all $R \ge R_\varepsilon$ and $\varphi \in C_0^\infty(\mathbb{R}^n)$. Now the definition of $\liminf_{|x|\to\infty} \Lambda_R(x;P)$ implies that

$$\int_{\mathbb{R}^n} (\Lambda_R(x,P) - \varepsilon)|\varphi(x)|^2 dx \geq (\liminf_{|x|\to\infty} (\Lambda_R(x,P) - 2\varepsilon)||\varphi||^2 \tag{2 16}$$

for all $\varphi \in C_0^\infty(\mathbb{R}^n \setminus \overline{B(0,R_0)})$ where R_0 is sufficiently large Therefore, (2 15) and (2 16) imply

$$\frac{(P\varphi,\varphi)}{||\varphi||^2} \geq \liminf_{|x|\to\infty} \Lambda_R(x,P) - 2\varepsilon$$

for $\varphi \in C_0^\infty(\mathbb{R}^n \setminus \overline{B(0,R_0)})$ Therefore

$$\Sigma(P) \geq \liminf_{|x|\to\infty} \Lambda_R(x,P) - 2\varepsilon \tag{2 17}$$

Now the only restriction on R is that $R \geq R_\varepsilon$ Therefore

$$\Sigma(P) \geq \lim_{R\to\infty} \liminf_{|x|\to\infty} \Lambda_R(x,P) - 2\varepsilon \tag{2 18}$$

But ε is arbitrary So (2 18) and (2 14) prove the lemma ▪

We now introduce a function $K(\omega) = K(\omega,P)$ defined on the sphere $S^{n-1} \subset \mathbb{R}^n$ which will allow us to estimate $\Lambda_R(x,P)$ for large $|x|$ by functions depending only on $x/|x|$ Roughly speaking $K(\omega)$ approximates the lower bound of the quadratic form $(P\varphi,\varphi)$ restricted to test functions φ which are supported in the intersection of a "small" neighborhood of infinity and a "thin" open cone in the direction of ω

2.6 Definition: *Let* $S^{n-1} = \{\omega \in \mathbb{R}^n \quad |\omega| = 1\}$ *For* $\omega \in S^{n-1}$, $0 < \varepsilon < \pi$ *and* $N > 0$ *define*

$$\Gamma_\omega^{\varepsilon,N} = \{x \in \mathbb{R}^n \quad <x,\omega> \, > |x|\cos\varepsilon \, , \, |x| > N\} \tag{2 19}$$

$$\Sigma^{\varepsilon,N}(\omega) = \inf\left\{\frac{(P\varphi,\varphi)}{||\varphi||^2} \quad \varphi \in C_0^\infty(\Gamma_\omega^{\varepsilon,N}) \, , \, \varphi \neq 0\right\} \tag{2 20}$$

$$K(\omega) = K(\omega,P) = \lim_{\varepsilon\to\infty} \lim_{N\to\infty} \Sigma^{\varepsilon,N}(\omega) \tag{2 21}$$

In (2 19) $< , >$ denotes the usual inner product in $\mathbb{R}^n$ $\Gamma_\omega^{\varepsilon,N}$ is a truncated cone with angle of opening ε The limit in (2 21) exists because $\Sigma^{\varepsilon,N}(\omega)$ increases as $N \to \infty$ and $\varepsilon \to 0$ Note the $K(\omega)$ may take the value $+\infty$

2.7 Lemma: *$K(\omega)$ is a lower semicontinuous function of ω on S^{n-1}. The following holds:*

$$\min\{K(\omega) : \omega \in S^{n-1}\} = \Sigma(P). \tag{2.22}$$

Proof: Let $\{\omega_j\}$ be a sequence of points in S^{n-1} such that $\lim_{j\to\infty} \omega_j = \omega$. Fix a number L such that $L < K(\omega)$. From the definition of $K(\omega)$ it follows that there exist $\varepsilon \in (0,\pi)$ and $N > 0$ such that $\Sigma^{\varepsilon,N}(\omega) > L$. Since $\omega_j \to \omega$ it is clear that $\Gamma_{\omega_j}^{\varepsilon/2,N} \subset \Gamma_{\omega}^{\varepsilon,N}$ for all $j \geq j_0$ for some j_0. Hence

$$K(\omega_j) \geq \Sigma^{\varepsilon/2,N}(\omega_j) \geq \Sigma^{\varepsilon,N}(\omega) > L$$

for $j \geq j_0$ which implies, since L is an arbitrary number $< K(\omega)$, that

$$\liminf_{j\to\infty} K(\omega_j) \geq K(\omega).$$

This proves that $K(\omega)$ is lower semicontinuous.

Next, to prove (2.22), fix $\delta > 0$ and write

$$\Sigma(P) = \sup_{K \text{ compact}} \inf\{(P\varphi,\varphi)/\ \|\varphi\|^2 : \varphi \in C_0^\infty(\mathbb{R}^n \setminus K)\ ,\ \varphi \neq 0\}$$
$$\leq \inf\{(P\varphi,\varphi)/\ \|\varphi\|^2 : \varphi \in C_0^\infty(\mathbb{R}^n \setminus K_0)\ ,\ \varphi \neq 0\} + \delta \tag{2.23}$$

for some compact set K_0.

If $R_0 > 0$ is chosen so that $K_0 \subset B(0;R_0)$, it follows from (2.23) that

$$\Sigma(P) \leq \inf\{(P\varphi,\varphi)/\ \|\varphi\|^2 : \varphi \in C_0^\infty(\Gamma_\omega^{\varepsilon,N})\ ,\ \varphi \neq 0\} + \delta$$
$$= \Sigma^{\varepsilon,N}(\omega) + \delta$$

for any $\omega \in S^{n-1}$, $0 < \varepsilon < \pi$ and $N > R_0$, which implies that

$$\Sigma(P) \leq \lim_{\varepsilon\to\infty} \lim_{N\to\infty} \Sigma^{\varepsilon,N}(\omega) + \delta = K(\omega) + \delta$$

for any $\omega \in S^{n-1}$. Hence

$$\Sigma(P) \leq \min\{K(\omega) : \omega \in S^{n-1}\}. \tag{2.24}$$

On the other hand using Lemma 2.5 it follows that given $\delta > 0$ there is an $R_1 > 0$ and a sequence of points $x_m \in \mathbb{R}^n$, $m = 1,2,\cdots$, with $|x_m| \to \infty$ so that

$$\Sigma(P) = \lim_{k\to\infty} \liminf_{|x|\to\infty} \Lambda_R(x;P)$$
$$\geq \liminf_{|x|\to\infty} \Lambda_{R_1}(x;P) - \delta/2$$
$$\geq \Lambda_{R_1}(x_m;P) - \delta \tag{2.25}$$

for $m = 1,2,\cdots$. We can assume without loss of generality that $(x_m/|x|) \to \omega_0$ for some $\omega_0 \in S^{n-1}$. Noting that $B(x_m;R_1) \subset \Gamma_{\omega_0}^{\varepsilon,N}$ for any given $0 < \varepsilon < \pi$ and $N > 0$ if m is sufficiently large, it follows from (2.25) that

$$\Sigma(P) \geq \Sigma^{\varepsilon,N}(\omega_0) - \delta$$

for any ε and N and thus

$$\Sigma(P) \geq \lim_{\varepsilon\to 0} \lim_{N\to\infty} \Sigma^{\varepsilon,N}(\omega_0) - \delta = K(\omega_0) - \delta$$
$$\geq \min\{K(\omega) : \omega \in S^{n-1}\} - \delta,$$

which together with (2.24), proves (2.22). ▪

Finally we prove a lemma which will let us obtain non-constant λ functions for use in Theorem 1.5 where the functions $\lambda(x)$ in question depend only on the direction of x. This lemma will be in particular useful in applications of Theorem 1.5 to eigenfunctions of multiparticle Schrödinger operators to be considered in Chapter 4.

2.8 Lemma: *Let $P = A + q$ satisfy the same conditions as before. Let $g(\omega)$ be a continuous function on S^{n-1} such that $g(\omega) < K(\omega) = K(\omega;P)$ for all $\omega \in S^{s-1}$. Then there exists $C > 0$ such that*

$$\int_{\Omega_C} (|\nabla_A \varphi(x)|^2 + q(x)|\varphi(x)|^2)dx \geq \int_{\Omega_C} g\left(\frac{x}{|x|}\right)|\varphi(x)|^2 dx \tag{2.26}$$

for all $\varphi \in C_0^\infty(\Omega_C)$ where $\Omega_C = \{x : |x| > C\}$.

Proof: Since $K(\omega) - g(\omega)$ is lower semicontinuous and positive we can choose $\delta > 0$ such that $g(\omega) + 2\delta < K(\omega)$ for all $\omega \in S^{n-1}$. By Lemma 2.3 there exists $R > 0$ such that

$$\int_{\mathbb{R}^n} (|\nabla_A \varphi(x)|^2 + q(x)|\varphi(x)|^2)dx \geq \int_{\mathbb{R}^n} (\Lambda_R(x;P) - \delta)|\varphi(x)|^2 dx$$

for every $\varphi \in C_0^\infty(\mathbb{R}^n)$. Therefore it is enough to show

$$\Lambda_R(x;P)-\delta \geq g\left(\frac{x}{|x|}\right) \tag{2.27}$$

for $|x| > C$. So fix $\omega_0 \in S^{n-1}$, $0 < \varepsilon_0 < \pi/2$ and $N_0 > 0$ so that

$$\Sigma^{\varepsilon_0,N_0}(\omega_0) > g(\omega_0) + 2\delta \tag{2.28}$$

and let U_0 be a neighborhood of ω_0 in S^{n-1} satisfying

$$g(\omega) < g(\omega_0) + \delta \quad \text{for} \quad \omega \in U_0, \tag{2.29}$$

$$\Gamma_\omega^{\varepsilon_0/2,N_0} \subset \Gamma_{\omega_0}^{\varepsilon_0,N_0} \quad \text{for} \quad \omega \in U_0. \tag{2.30}$$

Then (2.28) through (2.30) imply

$$\Sigma^{\varepsilon_0/2,N_0}(\omega) \geq \Sigma^{\varepsilon_0,N_0}(\omega_0) > g(\omega_0) + 2\delta > g(\omega) + \delta$$

for all $\omega \in U_0$. Now the compactness of S^{n-1} and a covering argument allows us to conclude that

$$\Sigma^{\varepsilon,N}(\omega) > g(\omega) + \delta$$

for every $\omega \in S^{n-1}$ and some $\varepsilon > 0$, $N > 0$. Finally let $C = \max\{N+R\,,\, R/\sin\varepsilon\}$. Then, if $|x| > C$

$$B(x;R) \subset \Gamma_{x/|x|}^{\varepsilon,N}$$

and therefore

$$\begin{aligned}\Lambda_R(x;P) &= \inf\left\{\frac{(P\varphi,\varphi)}{\|\varphi\|^2} : \varphi \in C_0^\infty(B(x;R))\,,\ \varphi \neq 0\right\}\\ &\geq \inf\left\{\frac{(P\varphi,\varphi)}{\|\varphi\|^2} : \varphi \in C_0^\infty(\Gamma_{x/|x|}^{\varepsilon,N})\,,\ \varphi \neq 0\right\}\\ &= \Sigma^{\varepsilon,N}\left(\frac{x}{|x|}\right)\\ &\geq g\left(\frac{x}{|x|}\right) + \delta,\end{aligned}$$

which proves (2.27) and completes the proof. ▪

Chapter 3 Self-Adjointness

In this chapter we consider the realization of $P = A+q$ as an operator on $L^2(\mathbb{R}^n)$. We show that when $\Sigma(P) > -\infty$ P together with a suitable domain define a self-adjoint operator H which is bounded from below. We also find that $\Lambda(P)$ and $\Sigma(P)$ are equal to the infimum of the spectrum and essential spectrum of H respectively. The results of this chapter will allow us to apply exponential decay results to eigenfunctions of H.

Throughout this chapter we assume that P satisfies the conditions imposed on $P = A+q$ in Chapter 2. That is, $A = -\sum_{i,j} \partial_j a^{ij} \partial_i$ where $a^{ij}(x)$ are continuous real valued bounded functions on $\mathbb{R}^n$ such that $[a^{ij}(x)]$ is positive for every x, and $q(x)$ is a real valued function in $L^1_{loc}(\mathbb{R}^n)$ such that $q_- \in M_{loc}(\mathbb{R}^n)$.

We begin by proving a lemma which will be needed in our study of the self-adjoint operator H.

3.1 Lemma: *Let $P = A+q$ be an elliptic operator on $\mathbb{R}^n$ verifying the conditions described above. There exists a positive continuous function $k(x)$ on $\mathbb{R}^n$ such that*

$$\int_{\mathbb{R}^n} k(x)(|\nabla_A \varphi|^2 + |q||\varphi|^2)dx \le (P\varphi,\varphi)_+ + ||\varphi||^2 \tag{3.1}$$

for every $\varphi \in C_0^\infty(\mathbb{R}^n)$.

Proof: We claim that it is enough to prove the lemma assuming the additional hypothesis $a^{ij} \in C^\infty(\mathbb{R}^n)$. For it follows from the approximation result in Appendix 1 that there are real valued functions $b^{ij} \in C^\infty(\mathbb{R}^n)$, $1 \le i,j \le n$, such that $[b^{ij}(x)]$ is a positive definite matrix satisfying

$$\frac{1}{2}[a^{ij}(x)] \le [b^{ij}(x)] \le 2[a^{ij}(x)] \tag{3.2}$$

where for matrices A and B, $A \ge B$ means that $A-B$ is positive semidefinite. Since $a^{ij} \in L^\infty(\mathbb{R}^n)$ it follows from (3.2) that $b^{ij} \in L^\infty(\mathbb{R}^n)$ for $1 \le i,j \le n$.

Now, suppose we have shown that there exists a positive continuous function $k_1(x)$ on $\mathbb{R}^n$ such that

$$\int_{\mathbb{R}^n} k_1(x)(|\nabla_B\varphi|^2 + 2|q||\varphi|^2)dx \le (P_1\varphi,\varphi)_+ + \|\varphi\|^2 \tag{3.3}$$

for all $\varphi \in C_0^\infty(\mathbb{R}^n)$ where

$$B(x,\partial) = -\sum_{i,j=1}^{n} \partial_j b^{ij}(x)\partial_i\,, \quad P_1 = B+2q.$$

Then we have from (3.2)

$$(P_1\varphi,\varphi)_+ \le 2(P\varphi,\varphi)_+ \tag{3.4}$$

while again from (3.2)

$$\int_{\mathbb{R}^n}\frac{1}{2}k_1(x)(|\nabla_A\varphi|^2 + |q||\varphi|^2)dx \le \int_{\mathbb{R}^n} k_1(x)(|\nabla_B\varphi|^2 + 2|q||\varphi|^2)dx \tag{3.5}$$

so that using (3.3) through (3.5), (3.1) follows with $k(x) = k_1(x)/4$. We thus assume in what follows that $a^{ij} \in C^\infty(\mathbb{R}^n)$.

Since $q_- \in M_{loc}(\mathbb{R}^n)$ we can apply Lemma 0.3 to conclude that there exists an increasing sequence $\{c_m\}$, $m = 0,1,2,\cdots$, $c_m \ge 1$, such that

$$\int_{\mathbb{R}^n} q_-|\varphi|^2dx \le \frac{1}{4}\int_{\mathbb{R}^n}|\nabla_A\varphi|^2 + c_m\int_{\mathbb{R}^n}|\varphi|^2dx$$

for each $\varphi \in C_0^\infty(B(0;m+2))$. Define a continuous function $c(y)$ of $|y|$ by setting $c(y) = c_m$ if $|y| = m$, $m = 0,1,2,\cdots$, and letting $c(y)$ be piecewise linear in $|y|$. Then $c(y) \ge 1$ and

$$\int_{\mathbb{R}^n} q_-|\varphi|^2dx \le \frac{1}{4}\int_{\mathbb{R}^n}|\nabla_A\varphi|^2dx + c(y)\int_{\mathbb{R}^n}|\varphi|^2dx \tag{3.6}$$

for all $\varphi \in C_0^\infty(B(y;1))$.

Using (3.6) and the definition of q_- we get that for $\varphi \in C_0^\infty(B(y;1))$

$$\int_{\mathbb{R}^n}(|\nabla_A\varphi|^2 + |q||\varphi|^2)dx = (P\varphi,\varphi) + 2\int_{\mathbb{R}^n} q_-|\varphi|^2dx$$

$$\le (P\varphi,\varphi)_+ + \frac{1}{2}\int_{\mathbb{R}^n}|\nabla_A\varphi|^2dx + 2c(y)\int_{\mathbb{R}^n}|\varphi|^2dx$$

$$\le (P\varphi,\varphi)_+ + \frac{1}{2}\int_{\mathbb{R}^n}(|\nabla_A\varphi|^2 + |q||\varphi|^2)dx + 2c(y)\int_{\mathbb{R}^n}|\varphi|^2dx.$$

Subtracting $\frac{1}{2}\int(|\nabla_A\varphi|^2 + |q||\varphi|^2)dx$ from both sides and dividing by $2c(y)$, noting that $c(y) \ge 1$, we obtain

$$(4c(y))^{-1}\int_{\mathbb{R}^n}(|\nabla_A\varphi|^2 + |q||\varphi|^2)dx \le (2c(y))^{-1}(P\varphi,\varphi)_+ + ||\varphi||^2$$

$$\le (P\varphi,\varphi)_+ + ||\varphi||^2 \qquad (3.7)$$

for all $\varphi \in C_0^\infty(B(y;1))$.

Let $\psi \in C_0^\infty(\mathbb{R}^n)$. We now manufacture a suitable function φ for use in (3.7). Let $\zeta \in C_0^\infty(\mathbb{R}^n)$ be a fixed real valued function such that

$$\zeta(x) = 0 \quad \text{if} \quad |x| \ge 1\ ,\ \zeta(x) = 1 \quad \text{if} \quad |x| \le \frac{1}{2},$$

$$\int_{\mathbb{R}^n}\zeta^2dx = 1.$$

Let $\zeta_y(x) = \zeta(x-y)$. Then $\zeta_y\psi \in C_0^\infty(B(y;1))$. Therefore, applying (3.7) to $\varphi = \zeta_y\psi$,

$$(4c(y))^{-1}\int_{\mathbb{R}^n}(|\nabla_A(\psi\zeta_y)|^2 + |q||\psi\zeta_y|^2)dx$$

$$\le (P(\psi\zeta_y),\psi\zeta_y)_+ + \int_{\mathbb{R}^n}|\zeta_y\psi|^2dx. \qquad (3.8)$$

On the other hand application of the identity (1.16) (see (1.16)') gives

$$(P(\psi\zeta_y),\psi\zeta_y) = \mathrm{Re}\int_{\mathbb{R}^n}(P\psi)\overline{\psi}\,\zeta_y^2dx + \int_{\mathbb{R}^n}|\nabla_A\zeta_y|^2|\psi|^2dx.$$

Setting $P\psi = f$ and noting that $|\nabla_A\zeta_y(x)|^2 \le D$ for some constant D not depending on x or y, it follows that

$$(P(\psi\zeta_y),\psi\zeta_y) \le \mathrm{Re}\int_{\mathbb{R}^n} f\,\overline{\psi}\zeta_y^2dx + D\int_{|x-y|<1}|\psi|^2dx. \qquad (3.9)$$

Define a positive continuous function

$$k_1(x) = \inf\{(4c(y))^{-1} : |x-y| \le 1\}.$$

Then (3.8) and (3.9) imply

$$\int_{|x-y|\le 1/2} k_1(x)(|\nabla_A\psi|^2 + |q||\psi|^2)dx$$
$$\le (P(\psi\zeta_y),\psi\zeta_y) + \int_{\mathbb{R}^n} |\psi|^2\zeta_y^2 dx \tag{3 10}$$
$$\le \mathrm{Re}\int_{\mathbb{R}^n} f\,\bar{\psi}\zeta_y^2 dx + D\int_{|x-y|<1} |\psi|^2 dx + \int_{\mathbb{R}^n} |\psi|^2\zeta_y^2 dx$$

In the first integral we have used the fact that $\zeta_y = 1$ when $|x-y| \le 1/2$ We now integrate the inequality (3 10) with respect to y (over $\mathbb{R}^n$) Letting V_r denote the volume of a ball of radius r in $\mathbb{R}^n$, we obtain

$$V_{1/2}\int_{\mathbb{R}^n} k_1(x)\,(|\nabla_A\psi|^2 + |q||\psi|^2)dx$$
$$\le \mathrm{Re}\int_{\mathbb{R}^n} f\,\bar{\psi}dx + DV_1\int_{\mathbb{R}^n} |\psi|^2 dx + \int_{\mathbb{R}^n} |\psi|^2 dx \tag{3 11}$$

where we have used that $\int_{\mathbb{R}^n} \zeta_y^2(x)dy = 1$ Recalling that $f = P\psi$ we obtain

$$V_{1/2}\int_{\mathbb{R}^n} k_1(x)(|\nabla_A\psi|^2 + |q||\psi|^2)dx$$
$$\le (P\psi,\psi) + (DV_1+1)\int_{\mathbb{R}^n} |\psi|^2 dx$$
$$\le (P\psi,\psi)_+ + (DV_1+1)\,||\psi||^2$$

which then gives

$$\int_{\mathbb{R}^n} V_{1/2}(DV_1+1)^{-1}\,k_1(x)(|\nabla_A\psi|^2 + |q||\psi|^2)dx$$
$$\le (DV_1+1)^{-1}(P\psi,\psi)_+ + ||\psi||^2 \le (P\psi,\psi)_+ + ||\psi||^2 \tag{3 12}$$

This proves the lemma with $k(x) = V_{1/2}(DV_1+1)^{-1}\,k_1(x)$ ▪

We now turn to the main result of this chapter

3 2 Theorem: *Let* $P = A+q = -\sum_{i,j}\partial_j a^{ij}(x)\partial_i + q(x)$ *be an elliptic operator on* $\mathbb{R}^n$ *satisfying the same conditions as in Lemma 3 1 Suppose also that* $\Sigma(P) > -\infty$ *Let* H be the operator on $L^2(\mathbb{R}^n)$, *with domain* $D(H)$ *defined as follows*

$$D(H) = \{u \quad u \in L^2(\mathbb{R}^n) \cap H^1_{loc}(\mathbb{R}^n),\ qu \in L^1_{loc}(\mathbb{R}^n)\ ,Au+qu \in L^2(\mathbb{R}^n)\} \tag{3 13}$$
$$Hu = Au + qu \quad \text{for } u \in D(H) \tag{3 14}$$

Then H is a self-adjoint operator bounded from below. Furthermore,

$$\inf \sigma(H) = \Lambda(P), \tag{3.15}$$

$$\inf \sigma_{ess}(H) = \Sigma(P), \tag{3.16}$$

where $\sigma(H)$ and $\sigma_{ess}(H)$ denote the spectrum and the essential spectrum of H respectively and $\Lambda(P)$ and $\Sigma(P)$ are as in Definition 2.4.

Remark 1: The action of the differential operator A on a function $u \in H^1_{loc}(\mathbb{R}^n)$ should be understood in the distribution sense (see (1.6)). Thus writing in (3.13) that $Au + qu \in L^2(\mathbb{R}^n)$ means that there exists a (necessarily unique) function $f \in L^2(\mathbb{R}^n)$ such that $Au + qu = f$ in the weak sense (1.5) (with $\Omega = \mathbb{R}^n$).

Remark 2: The domain $D(H)$ defined by (3.13) is the maximal domain of functions in $L^2(\mathbb{R}^n) \cap H^1_{loc}(\mathbb{R}^n)$ on which the differential operator P acts in a natural way as an unbounded linear operator in $L^2(\mathbb{R}^n)$. The theorem shows that H is the *unique* self-adjoint realization of P in $L^2(\mathbb{R}^n)$ with domain contained in $L^2(\mathbb{R}^n) \cap H^1_{loc}(\mathbb{R}^n)$.

Remark 3: The proof of the theorem will show that if $u \in D(H)$ then $|q|^{1/2}u \in L^1_{loc}(\mathbb{R}^n)$. This will imply the following alternative characterization of the domain of H,

$$D(H) = \{u : u \in L^2(\mathbb{R}^n) \cap H^1_{loc}(\mathbb{R}^n)\,,\ |q|^{1/2}u \in L^1_{loc}(\mathbb{R}^n)\,,\ Au+qu \in L^2(\mathbb{R}^n)\}.$$

Proof of Thoerem 3.2: We start with some observations. Introduce in $\mathbb{R}^n$ the Riemannian metric

$$ds_1^2 = \sum_{i,j=1}^{n} a_{ij}(x)dx_i dx_j$$

where the matrix $[a_{ij}(x)]$ is the inverse of $[a^{ij}(x)]$. Since by assumption the coefficients $a^{ij}(x)$ of A are bounded on $\mathbb{R}^n$ it follows that $[a_{ij}(x)] \geq \delta I$ (I the identity matrix) for some constant $\delta > 0$ and all $x \in \mathbb{R}^n$. Hence, if $\rho_1(x,y)$ denotes the geodesic distance between x, y in the Riemannian metric ds_1^2 we have: $\rho_1(x,y) \geq \delta|x-y|$ which implies that $\mathbb{R}^n$ is complete in the metric ρ_1.

Next we observe that without loss of generality we may assume that P verifies the condition

$$(P\varphi,\varphi) \geq \|\varphi\|^2 \quad \text{for} \quad \varphi \in C_0^\infty(\mathbb{R}^n). \tag{3.17}$$

Indeed, Lemma 2.3 implies that for R larger than some R_ε,

$$(P\varphi,\varphi) \geq \int_{\mathbb{R}^n} (\Lambda_R(x;P)-\varepsilon)|\varphi(x)|^2 dx. \tag{3.18}$$

Since, by Lemma 2.5

$$\lim_{R\to\infty} \liminf_{|x|\to\infty} \Lambda_R(x;P) = \Sigma(P) > -\infty$$

and $\liminf_{|x|\to\infty} \Lambda_R(x;P)$ is a non-increasing function of R, we have

$$\liminf_{|x|\to\infty} \Lambda_R(x;P) \geq \Sigma(P)$$

for every R. Thus since $\Lambda_R(x;P)$ is also continuous there exists a constant C such that

$$\Lambda_R(x;P) - \varepsilon \geq C \quad \text{for every } x \in R^n.$$

Together with (3.18) this implies that

$$(P\varphi,\varphi) \geq C\|\varphi\|^2 \quad \text{for all} \quad \varphi \in C_0^\infty(\mathbb{R}^n). \tag{3.19}$$

Now we add a constant γ to P so that (3.19) holds for $P+\gamma$ with $C \geq 1$ and note that if we can prove the theorem for $H_\gamma = H+\gamma$ then it is also true for H.

We thus assume that (3.17) holds, observing that this implies that the λ-Condition of Theorem 1.5 is satisfied with $\lambda \equiv 1$. We now apply the theorem to the function $u \in D(H)$ which we consider as a solution of the differential equation $Pu = f$ with $f = Hu$. We use the theorem with $h \equiv 0$ (all other hypotheses are easily checked). Since as we have seen before $\mathbb{R}^n$ is complete in the metric ρ_1 it follows from (1.14) that

$$\|u\| \leq \|Hu\| \quad \text{for every} \quad u \in D(H) \tag{3.20}$$

which says that H is injective.

Our strategy now is to find a self-adjoint inverse for H. Then the self-adjointness of H will follow.

Define

$$|||\varphi|||^2 = (P\varphi,\varphi) \quad \text{for} \quad \varphi \in C_0^\infty(\mathbf{R}^n). \tag{3.21}$$

Then it follows from Lemma 3.1 and (3.17) that there exists a positive continuous function $k(x)$ on $\mathbf{R}^n$ such that

$$2|||\varphi|||^2 \geq (P\varphi,\varphi) + ||\varphi||^2 \geq \int_{\mathbf{R}^n} k(x)(|\nabla_A\varphi|^2 + |q||\varphi|^2)dx$$

for every $\varphi \in C_0^\infty(\mathbf{R}^n)$. Adding th.s inequality to (3.17) we obtain

$$|||\varphi|||^2 \geq \frac{1}{3}||\varphi||^2 + \frac{1}{3}\int_{\mathbf{R}^n} k(x)(|\nabla_A\varphi|^2 + |q||\varphi|^2)dx \tag{3.22}$$

for every $\varphi \in C_0^\infty(\mathbf{R}^n)$. Let V be the completion of $C_0^\infty(\mathbf{R}^n)$ with respect to the Hilbert space norm $|||\cdot|||$. Denote by $(\cdot,\cdot)_V$ the scalar product in the Hilbert space V. Note that in view of (3.22) we have

$$V \subset \{u : u \in L^2(\mathbf{R}^n) \cap H^1_{loc}(\mathbf{R}^n)\,,\ |q|^{1/2}u \in L^2_{loc}(\mathbf{R}^n)\} \tag{3.23}$$

In particular $V \subset L^2(\mathbf{R}^n)$. Also by (3.17) $||u|| \leq |||u|||$ for $u \in V$.

Given $f \in L^2(\mathbf{R}^n)$ consider the map

$$\Phi_f(u) = (u,f)_{L^2} \quad \text{for} \quad u \in V.$$

Then Φ_f is a bounded linear functional on V since

$$|\Phi_f(u)| = |(u,f)_{L^2}| \leq ||u||\,||f|| \leq |||u|||\,||f||.$$

Therefore by Riesz's representation theorem there exists a unique $v_f \in V$ such that

$$\Phi_f(u) = (u,f)_{L^2} = (u,v_f)_V \tag{3.24}$$

for every $u \in V$. Define $T : L^2(\mathbf{R}^n) \to V \subset L^2(\mathbf{R}^n)$ by $Tf = v_f$. Then for $f \in L^2(\mathbf{R}^n)$ with $||f|| = 1$

$$\begin{aligned} ||Tf||^2 &\leq |||Tf|||^2 = (v_f,v_f)_V = (v_f,f)_{L^2} \\ &= (Tf,f)_{L^2} \leq ||Tf|| \end{aligned}$$

which implies $||Tf|| \leq 1$ and $(Tf,f)_{L^2} \geq 0$. Thus T is a self-adjoint operator with $\sigma(T) \subset [0,1]$ and Ran $T \subset V$.

Next, recalling (3.23) and the definition of Tf it follows from the fact that $C_0^\infty(\mathbf{R}^n)$ is dense in V (in the norm $|||\cdot|||$) that

$$(\varphi,f)_{L^2} = (\varphi,Tf)_V = \int_{\mathbf{R}^n}(\nabla_A\varphi\cdot\overline{\nabla_A Tf} + q\varphi\overline{Tf})dx$$

for every $\varphi \in C_0^\infty(\mathbf{R}^n)$ and for any given $f \in L^2(\mathbf{R}^n)$. This shows that the function $u = Tf \in V$ verifies the equation $Pu = f$ in the usual weak quadratic form sense. It thus follows that $Tf \in D(H)$ and that $HTf = f$. Thus we have shown that H is a one to one map from $D(H)$ onto $L^2(\mathbf{R}^n)$ such that $H^{-1} = T$ is self-adjoint. This proves that H is self-adjoint. Since $\sigma(T) \subset [0,1]$ it follows that $\sigma(H) \subset [1,\infty)$.

The above considerations show that $D(H) = \text{Ran } T \subset V$. Note that, in view of (3.23) this implies that $|q|^{1/2}u \in L^1_{loc}(\mathbf{R}^n)$ if $u \in D(H)$. We next observe that $D(H)$ is dense in the Hilbert space V. Indeed if this were not the case there would exist an element $v_0 \in V$, $v_0 \neq 0$, such that $(v_0,Tf)_V = 0$ for every $f \in L^2(\mathbf{R}^n)$. In view of (3.24) this would imply that $(v_0,f)_{L^2} = 0$ for every $f \in L^2(\mathbf{R}^n)$ which contradicts $v_0 \neq 0$.

Consider the operator $H^{1/2}$. We claim that $D(H^{1/2}) = V$. Indeed it follows from the spectral theorem that $D(H^{1/2})$ = closure of $D(H)$ in the graph norm $(||u||^2 + (Hu,u))^{1/2}$ which in our case is equivalent to the norm $(Hu,u)^{1/2} = |||u|||$ (we are using here the relation $(f,Tf)^{1/2} = |||Tf|||$). Since we have just seen that the closure of $D(H)$ in the norm $|||\cdot|||$ is V the result follows.

By the above remarks and the fact that $C_0^\infty(\mathbf{R}^n)$ is dense in the Hilbert space V, we get

$$\inf \sigma(H) = \inf_{u\in D(H)} \frac{(Hu,u)}{||u||^2} = \inf_{u\in V} \frac{|||u|||^2}{||u||^2}$$

$$= \inf_{\varphi\in C_0^\infty(\mathbf{R}^n)} \frac{|||\varphi|||^2}{||\varphi||^2} = \inf_{\varphi\in C_0^\infty(\mathbf{R}^n)} \frac{(P\varphi,\varphi)}{||\varphi||^2} = \Lambda(P).$$

This proves (3.15).

To prove (3.16) we use Lemma 2.3 again. It follows from the Lemma that given $\varepsilon > 0$ there exists a number $R > 0$ such that

$$(P\varphi,\varphi) \geq \int_{\mathbb{R}^n} (\Lambda_R(x;P) - \frac{\varepsilon}{2}) |\varphi|^2 dx$$

for every $\varphi \in C_0^\infty(\mathbb{R}^n)$. Since $\liminf_{|x| \to \infty} \Lambda_R(x;P) \geq \Sigma(P)$, it follows that

$$\Lambda_R(x;P) \geq \Sigma(P) - \frac{\varepsilon}{2} \quad \text{for} \quad |x| \geq a$$

for some $a > 0$ sufficiently large. Since as a function of x $\Lambda_R(x;P)$ is locally bounded, we have

$$\Lambda_R(x;P) \geq \Sigma(P) - C \quad \text{for} \quad |x| \leq a$$

for some constant C. Choose a non negative function $\chi \in C_0^\infty(\mathbb{R}^n)$ such that $\chi(x) \geq C$ for $|x| \leq a$. Set $P_\chi = P + \chi$. Then it is clear from the above that

$$(P_\chi \varphi,\varphi) \geq (\Sigma(P) - \varepsilon) \int_{\mathbb{R}^n} |\varphi|^2 dx$$

for every $\varphi \in C_0^\infty(\mathbb{R}^n)$.

Introduce now the multiplication operator $\chi : u \to \chi u$ which is a bounded self-adjoint operator in $L^2(\mathbb{R}^n)$ and consider the operator $H_\chi = H + \chi$ with $D(H_\chi) = D(H)$. Then H_χ is self-adjoint and previous considerations applied to H_χ show that

$$\inf \sigma(H_\chi) = \inf_{\varphi \in C_0^\infty(\mathbb{R}^n)} \frac{((P+\chi)\varphi,\varphi)}{\|\varphi\|^2} \geq \Sigma(P) - \varepsilon. \tag{3.25}$$

Observe now that the operator χ is H-compact. To show this it suffices to demonstrate that $\chi H^{-1} = \chi T$ is a compact operator, i.e. that $Y = \chi T \{ f \in L^2(\mathbb{R}^n) : \|f\| \leq 1 \}$ has compact closure. Since every $u \in Y$ belongs to $H^1_{loc}(\mathbb{R}^n)$ and has support in some fixed ball B containing supp χ, it suffices to show that $\|u\| + \sum_{i=1}^n \|\partial_i u\| \leq c$ for every $u \in Y$ and some constant c. For by Rellich's theorem [3; Theorem 3.8, p. 30] the closure of

$$\{ u \in H^1_{loc}(\mathbb{R}^n) : \text{supp } u \subset B, \ \|u\| + \sum_{i=1}^n \|\partial_i u\| \leq c \}$$

is compact in $L^2(\mathbb{R}^n)$.

Suppose $u = \chi Tf$ with $\|f\| \le 1$ Then

$$\|u\| = \|\chi Tf\| \le \|\chi\|_{L^\infty(\mathbb{R}^n)} \|T\| \|f\| \le c_1$$

We now compute

$$\begin{aligned}\|\partial_i u\| &\le \|(\partial_i \chi) Tf\| + \|\chi(\partial_i Tf)\| \\ &\le c_2 + c_3 \{\int_B \sum_{j=1}^{n} (\partial_j Tf)^2 dx\}^{1/2} \\ &\le c_2 + c_4 \{\int_B (|\nabla_A Tf|^2 + |q|\,|Tf|^2) dx\}^{1/2} \\ &\le c_2 + c_5 |||Tf||| \qquad (3\ 26) \\ &= c_2 + c_5 (Tf, f)_{L^2}^{1/2} \\ &\le c_2 + c_5\end{aligned}$$

where we have used the lower bound $[a^{ij}(x)] \ge \delta I$ for some constant $\delta > 0$ for all $x \in B$ to derive the third inequality, the inequality (3 22) (which by density holds for $\varphi = Tf$) to derive the fourth inequality and the relations

$$|||Tf|||^2 = (Tf, f)_{L^2} \le \|T\| \|f\|^2 \le 1$$

to derive the fifth and sixth inequalities in (3 26) It thus follows that χ is H-compact

Hence by Weyl's theorem [35, Theorem XIII 14, p 112] it follows that

$$\sigma_{ess}(H) = \sigma_{ess}(H_\chi) \qquad (3\ 27)$$

Combining (3 27) with (3 25) we get

$$\inf \sigma_{ess}(H) \ge \Sigma(P) - \varepsilon$$

and since ε was arbitrary it follows that

$$\inf \sigma_{ess}(H) \ge \Sigma(P) \qquad (3\ 28)$$

Finally we show the reverse inequality To this end we choose any positive number μ such that $\mu < \inf \sigma_{ess}(H)$ Let $E(\mu) = E((\mu, \infty))$ be the spectral projection of H which corresponds to the interval (μ, ∞) Then $I - E(\mu)$ is a finite rank projection, so $I - E(\mu) = \sum_{i=1}^{N} (\ ,\psi_i)\psi_i$ for $\psi_i \in D(H)$ Thus if φ is a test function with support outside $B(0;R)$ we have, using Schwarz's inequality

$$|||(I-E(\mu))\varphi||| \le \sum_{i=1}^{N} |(\varphi,\psi_i)| \; |||\psi_i|||$$

$$\le \sum_{i=1}^{N} \{ \int_{|x|>R} |\psi_i|^2 dx \}^{1/2} \; |||\psi_i||| \; ||\varphi|| .$$

Therefore given $\varepsilon > 0$ there is an R_ε so that

$$|||(I-E(\mu))\varphi||| \le \varepsilon ||\varphi|| \tag{3.29}$$

for all $\varphi \in C_0^\infty(\mathbb{R}^n \setminus \overline{B(0;R_\varepsilon)})$. From the spectral theorem we have

$$|||E(\mu)\varphi|||^2 = \int_\mu^\infty \lambda d(\varphi, E((-\infty,\lambda))\varphi)$$

$$\ge \mu ||E(\mu)\varphi||^2. \tag{3.30}$$

(We are using throughout that $|||u||| = (Hu,u)^{1/2}$ for $u \in V = D(H^{1/2})$). Combining (3.29) and (3.30), recalling that $||\cdot|| \le |||\cdot|||$, we have

$$(P\varphi,\varphi)^{1/2} = |||\varphi||| \ge |||E(\mu)\varphi||| - |||(I-E(\mu))\varphi|||$$

$$\ge \mu^{1/2}||E(\mu)\varphi|| - \varepsilon||\varphi|| \ge \mu^{1/2}||\varphi|| - \mu^{1/2}||(I-E(\mu)\varphi|| - \varepsilon||\varphi||$$

$$\ge (\mu^{1/2}-\varepsilon)||\varphi|| - \mu^{1/2}|||(I-E(\mu))\varphi|||$$

$$\ge (\mu^{1/2}-\varepsilon(\mu^{1/2}+1))||\varphi|| \tag{3.31}$$

for all $\varphi \in C_0^\infty(\mathbb{R}^n \setminus \overline{B(0;R)})$.

Thus using (3.31) we find that

$$\Sigma(P) = \sup_{K \text{ compact}} \inf\{ \frac{(P\varphi,\varphi)}{||\varphi||^2} : \varphi \in C_0^\infty(\mathbb{R}^n \setminus K) \, , \, \varphi \ne 0 \}$$

$$\ge \inf\{ \frac{(P\varphi,\varphi)}{||\varphi||^2} : \varphi \in C_0^\infty(\mathbb{R}^n \setminus \overline{B(0;R_\varepsilon)} \, , \, \varphi \ne 0 \}$$

$$\ge (\mu^{1/2}-\varepsilon(\mu^{1/2}+1))^2$$

for $\varepsilon > 0$ sufficiently small. Letting $\varepsilon \to 0$ we find $\Sigma(P) \ge \mu$. Finally since μ was an arbitrary positive number less than $\inf \sigma_{ess}(H)$ we get $\Sigma(P) \ge \inf \sigma_{ess}(H)$. This together with (3.28) proves (3.16) and completes the proof of the theorem. ▪

Chapter 4 L^2 Exponential Decay

Applications to eigenfunctions of N-body Schrodinger Operators

In this chapter we apply Theorem 1 5 to prove L^2 exponential decay results In Theorem 4 1 we consider the general class of elliptic operators $A+q$ studied in Chapter 2 and Chapter 3 without taking into account in a detailed way the variation of $\Lambda(x,R)$ in distant regions of $\mathbf{R}^n$ Thus replacing Λ by a constant we prove the theorem using a metric ρ_λ with $\lambda =$ constant In Theorem 4 4 we consider operators $A+q$ with constant a^{ij} but take into consideration the variation of $\Lambda(x,R)$ with direction for large $|x|$ by making use of the function $K(\omega)$ introduced in Chapter 2 We thus obtain an L^2 exponential decay result for solutions of $Au+qu = zu$ at a neighborhood of infinity which involve a metric ρ_c with c discontinuous but still lower semicontinuous The use of such metrics requires an additional technical result given in Lemma 4 3 We then consider decay problems for eigenfunctions of Schrodinger operators $P = -\Delta+V$ for a general class of multiparticle type potentials V For such operators the function $K(\omega)$ can be described explicitly and takes a relatively simple form Combining this information with Theorem 4 4 we obtain in Theorem 4 9 a general exponential decay result We conclude this chapter with a discussion of the N-body problem The exponential decay estimates for eigenfunctions of the corresponding Schrodinger operator are obtained as a special case of Theorem 4 9

Throughout the chapter we use the results of Chapter 3 to apply the exponential decay theorems to L^2 eigenfunctions of the self-adjoint realization of P

4.1 Theorem: *Let $P = A+q = -\sum_{i,j}\partial_j a^{ij}(x)\partial_i + q(x)$ be an elliptic operator on $\mathbf{R}^n$ satisfying the hypotheses of Theorem 3 2 Let $\rho(x,y)$ be the geodesic distance from x to y in $\mathbf{R}^n$ in the Riemannian metric*

$$ds^2 = \sum_{i,j} a_{ij}(x)\,dx_i\,dx_j$$

where $[a_{ij}(x)]$ is the inverse of the matrix $[a^{ij}(x)]$ Set $\rho(x) = \rho(x,x^0)$ where

x^0 is some fixed point in $\mathbb{R}^n$. Let $\Omega = \mathbb{R}^n \setminus K$ where K is some compact set and let $z \in \mathbb{C}$ with Re z *$= \mu < \Sigma$ where $\Sigma = \Sigma(P)$ is defined by (2.12).*

Let $\psi(x)$ be a solution of the equation $P\psi = z\psi$ in Ω in the sense that $\psi \in H^1_{loc}(\Omega)$, $q\psi \in L^1_{loc}(\Omega)$, and

$$\int_\Omega (\nabla_A \psi \cdot \nabla_A \varphi + q\psi\varphi)dx = \int_\Omega z\psi\varphi dx$$

for every $\varphi \in C_0^\infty(\Omega)$. Suppose that

$$\int_\Omega |\psi(x)|^2 e^{-2\beta\rho(x)} dx < \infty \tag{4.1}$$

for some $\beta < (\Sigma-\mu)^{1/2}$, $\beta \geq 0$. Then

$$\int_\Omega |\psi(x)|^2 e^{2\alpha\rho(x)} dx < \infty \tag{4.2}$$

for any $\alpha < (\Sigma-\mu)^{1/2}$.

Proof: Given $0 < \alpha < (\Sigma-\mu)^{1/2}$ choose $\varepsilon > 0$ such that $\alpha < (\Sigma-\mu-\varepsilon)^{1/2}$ and $\beta < (\Sigma-\mu-\varepsilon)^{1/2}$. It follows from the definition of Σ that there exists a number $R_\varepsilon > 0$ such that setting $\Omega_\varepsilon = \{x : |x| > R_\varepsilon\}$, we have

$$\int_{\Omega_\varepsilon} (|\nabla_A \varphi|^2 + q|\varphi|^2)dx \geq (\Sigma-\varepsilon) \int_{\Omega_\varepsilon} |\varphi|^2 dx \tag{4.3}$$

for all $\varphi \in C_0^\infty(\Omega_\varepsilon)$. Without loss of generality we may assume that $\overline{\Omega}_\varepsilon \subset \Omega$. By subtracting μ from each side of (4.3) we obtain

$$\text{Re} \int_{\Omega_\varepsilon} (|\nabla_A \varphi|^2 + (q-z)|\varphi|^2)dx \geq (\Sigma-\mu-\varepsilon) \int_{\Omega_\varepsilon} |\varphi|^2 dx.$$

This is the λ-Condition (1.12) of Theorem 1.5 where the function q of the theorem is the present $q-z$ and $\lambda(x) \equiv \Sigma-\mu-\varepsilon$. We now check the remaining hypotheses.

Clearly $(q-z) \in L^1_{loc}(\Omega_\varepsilon)$ and $(q-z)_- \in M_{loc}(\Omega_\varepsilon)$ by our assumptions on q. Let $h(x) = \alpha\,\rho(x)$. Then, using Theorem 1.4,

$$|\nabla_A h(x)|^2 = \alpha^2 |\nabla_A \rho(x)|^2 \leq \alpha^2 < \lambda \quad a.e. \tag{4.4}$$

Take in Theorem 1.5 $f \equiv 0$ and $u = \psi$ which in view of our assumptions is a solution of

$$A(x,\partial)u + (q-z)u = 0$$

in Ω_ε. Finally we need to show that

$$\int_{\Omega_\varepsilon} |\psi(x)|^2 \lambda e^{-2(1-\delta)\rho_\lambda(x)} dx < \infty \tag{4.5}$$

for some $\delta > 0$ where for some fixed $x^1 \in \Omega_\varepsilon$

$$\rho_\lambda(x) = \rho_\lambda(x,x^1) = \inf_\gamma \int_0^1 \lambda^{1/2} [\sum_{i,j} a_{ij}(\gamma(t))\dot\gamma_i(t)\dot\gamma_j(t)]^{1/2} dt$$

and the infimum is taken over all absolutely continuous paths $\gamma : [0,1] \to \Omega_\varepsilon$ joining x^1 to x. Clearly $\lambda^{1/2}\rho(x,x^1) \le \rho_\lambda(x,x^1)$ so that by the triangle inequality

$$\lambda^{1/2}\rho(x) \le \rho_\lambda(x) + c$$

where $c = \lambda^{1/2}\rho(x^1,x^0)$. We now choose $\delta = 1 - \beta\lambda^{-1/2}$. Then $0 < \delta < 1$ and $(1-\delta)\rho_\lambda(x) \ge \beta\,\rho(x) - \text{constant}$ so that (4.5) follows from our assumption (4.1).

Therefore we can apply Theorem 1.5 to obtain

$$\int_{\Omega_{\varepsilon,d}} |\psi|^2(\lambda - |\nabla_A h|^2) e^{2h}\, dx$$

$$\le \frac{2(1+2d)}{d^2} \int_{\Omega_\varepsilon \setminus \Omega_{\varepsilon,d}} \lambda |\psi|^2 e^{2h}\, dx$$

for any $d > 0$ where $\Omega_{\varepsilon,d} = \{x : x \in \Omega_\varepsilon\,,\, \rho_\lambda(x,\partial\Omega_\varepsilon) > d\}$. From (4.4) it follows that

$$\lambda - \alpha^2 \le (\lambda - |\nabla_A h|^2)$$

so that recalling that $h(x) = \alpha\rho(x)$ we have

$$\int_{\Omega_{\varepsilon,d}} |\psi(x)|^2 e^{2\alpha\rho(x)} dx \le C_d(\lambda-\alpha^2)^{-1} \int_{\Omega_\varepsilon\setminus\Omega_{\varepsilon,d}} |\psi(x)|^2 e^{2\alpha\rho(x)} dx \tag{4.6}$$

where $C_d = 2(1+2d)d^{-2}$.

Since the a^{ij} were assumed bounded we have, as before, that $\rho_\lambda(x,y) \ge$ Const.$|x-y|$ where $|x-y|$ is the Euclidean distance. This implies that $\Omega_\varepsilon \setminus \Omega_{\varepsilon,d}$ is contained in a compact subset of Ω. Thus, since $\psi \in L^2_{loc}(\Omega)$ the right side of (4.6) is finite. Finally the assumption (4.1) implies that

$$\int_{\Omega\setminus\Omega_{\epsilon,d}} |\psi(x)|^2 e^{2a\rho(x)}dx < \infty.$$

Therefore we conclude

$$\int_{\Omega} |\psi(x)|^2 e^{2a\rho(x)}dx < \infty$$

and the proof is complete. ▪

We now apply this result to L^2 eigenfunctions.

4.2 Corollary: *Let $P = A+q$ be an elliptic operator on $\mathbb{R}^n$ satisfying the same conditions as in Theorem 3.2. Let H be the self-adjoint realization of P in $L^2(\mathbb{R}^n)$ defined in Theorem 3.2. Let ψ be an eigenfunction of H with eigenvalue $\mu < \inf \sigma_{ess}(H)$. Then*

$$\int_{\mathbb{R}^n} |\psi|^2 \, e^{2a\rho(x)}dx < \infty \tag{4.7}$$

for every $a < (\Sigma-\mu)^{1/2}$ where $\rho(x)$ is defined as in Theorem 4.1.

Proof: By Theorem 3.2 $\Sigma = \inf \sigma_{ess}(H)$. Since $\psi \in L^2(\mathbb{R}^n)$ we apply Theorem 4.1 with $\beta = 0$. ▪

Note that if $\sigma_{ess}(H)$ is empty then (4.7) holds for any $a \in \mathbb{R}$.

In the following theorems we will have to consider distance functions ρ_λ defined by a Riemannian metric

$$ds_\lambda^2 = \lambda(x) \sum_{i,j=1}^{n} a_{ij}\, dx_i\, dx_j$$

where $\lambda(x)$ is lower semicontinuous but not necessarily continuous function and where $[a_{ij}]$ is a constant positive definite matrix. As before we define

$$\rho_\lambda(x,y) = \inf_{\gamma} \int_0^1 [\lambda(\gamma(t)) \sum_{i,j=1}^{n} a_{ij}\dot\gamma_i(t)\dot\gamma_j(t)]^{1/2} dt \tag{4.8}$$

where the infimum is taken over all absolutely continuous paths $\gamma : [0,1] \to \mathbb{R}^n$ which join y to x. The following convergence result for a sequence of metrics ρ_{λ_j} with $\lambda_j \uparrow \lambda$ will be needed later on.

4.3 Lemma *Let $[a_{ij}]$ be a constant positive definite matrix. Let $\{c_j(x)\}$, $j = 1,2,\ldots$, be a non-decreasing sequence of non-negative, locally bounded lower semicontinuous functions on $\mathbf{R}^n$, such that $c_j(x) \geq \delta > 0$, except for possibly finitely many x. Set $c(x) = \lim_{j\to\infty} c_j(x)$ and suppose that $c(x)$ is locally bounded. Then $c(x)$ is lower semicontinuous and*

$$\lim_{j\to\infty} \rho_{c_j}(x,y) = \rho_c(x,y) \tag{4.9}$$

uniformly in (x,y) in any compact subset of $\mathbf{R}^n \times \mathbf{R}^n$.

Proof. We begin by showing that if λ is a lower semicontinuous function then $\rho_\lambda(x,y)$ does not change when we change the value of λ at finitely many points. Clearly it is enough to show this for one point. Suppose $\lambda(x) = \tilde{\lambda}(x)$ except at x^0. Then since for any absolutely continous path γ, $\dot\gamma(t) = 0$ a.e. on $\gamma^{-1}(\{x^0\})$ (see Lemma 5.5) we have $\lambda(\gamma(t))|\dot\gamma(t)|^2 = \tilde{\lambda}(\gamma(t))|\dot\gamma(t)|^2$ a.e. for $t \in [0,1]$. Thus it follows that $\rho_\lambda = \rho_{\tilde\lambda}$. Therefore we can assume without loss of generality that $c_j(x) \geq \delta > 0$ everywhere on $\mathbf{R}^n$.

Since the c_j are non-decreasing, the ρ_{c_j} are a non-decreasing sequence of continuous functions on $\mathbf{R}^n \times \mathbf{R}^n$. It suffices, therefore, to prove pointwise convergence since by Dini's theorem this implies the uniform convergence of (4.9) on compact sets. Clearly,

$$\lim_{j\to\infty} \rho_{c_j}(x,y) \leq \rho_c(x,y) \tag{4.10}$$

We now prove the reverse inequality and thus establish the lemma. We first prove this under the additional assumption that the $c_j(x)$ are continuous. Without loss of generality we assume that the matrix $[a_{ij}]$ is the identity matrix, for by making the non-singular transformation $x \to [a_{ij}]^{1/2}x$ we can reduce the situation to this case. Therefore we assume $ds^2 = \Sigma a_{ij}\,dx_i\,dx_j$ is the Euclidean Reimannian metric.

Let $L_j(\gamma)$ denote the length of the path γ in the metric $ds_{c_j}^2 = c_j ds^2$ and let $\ell(\gamma)$ denote the Euclidean length in the metric ds^2. Fix two distinct points

$x,y \in \mathbb{R}^n$. The definition of the $\rho_{c_j}(x,y)$ implies that we can find a sequence of absolutely continuous paths $\gamma_j : [0,1] \to \mathbb{R}^n$ joining y to x such that

$$L_j(\gamma_j) < \rho_{c_j}(x,y) + \frac{1}{j},\ j = 1,2,\cdots. \tag{4.11}$$

Moreover, since the $c_j(x)$ are assumed to be continuous it follows by a standard approximation argument that the paths γ_j in (4.11) can be chosen as non self-intersecting polygonal lines. We shall reparametrize γ_j by choosing $s/\ell(\gamma_j)$ as the new parameter where s denotes the Euclidean arc length parameter on γ_j measured from x. With this choice $t \to \gamma_j(t)$ is a piecewise linear function on $[0,1]$ such that $|\dot{\gamma}_j(t)| = \ell(\gamma_j)$.

We want to apply Ascoli's theorem to the family of paths γ_j. Since $c_j(x) \geq \delta > 0$

$$\ell(\gamma_j) \leq \delta^{-1/2} L_j(\gamma_j) \leq \delta^{-1/2}(\rho_c(x,y)+1). \tag{4.12}$$

Therefore

$$\begin{aligned} \max\{|\gamma_j(t)| : 0 \leq t \leq 1\} &\leq |x| + \ell(\gamma_j) \\ &\leq K(x,y) \end{aligned} \tag{4.13}$$

where $K(x,y)$ does not depend on j. By (4.13) the sequence $\{\gamma_j\}$ is uniformly bounded. It is also equicontinuous since

$$\begin{aligned} |\gamma_j(t_1) - \gamma_j(t_2)| &= |\int_{t_1}^{t_2} \dot{\gamma}_j(t)dt| \\ &\leq \ell(\gamma_j)|t_1-t_2| \\ &\leq \delta^{-1/2}(\rho_c(x,y)+1)|t_1-t_2| \end{aligned} \tag{4.14}$$

where the last inequality follows from (4.12). Therefore Ascoli's theorem applies. By passing to a subsequence we therefore can assume that there is a continuous path $\gamma(t)$ such that $\gamma_j(t) \to \gamma(t)$ uniformly on $[0,1]$. Since the $\ell(\gamma_j)$ are bounded we can choose this subsequence so that $\ell(\gamma_j) \to \ell$ for some number ℓ. Taking the limit as $j \to \infty$ of the second line of (4.14) we obtain

$$|\gamma(t_1) - \gamma(t_2)| \leq \ell\, |t_1-t_2| \tag{4.15}$$

Hence $\gamma : t \to \gamma(t)$, $t \in [0,1]$, defines an absolutely continuous path joining y to x. Note also that in view of (4.15) we have

$$|\dot{\gamma}(t)| \le \ell \quad a.e. \quad in \quad [0,1]. \tag{4.16}$$

Now let i be fixed and $j \ge i$. Then

$$\begin{aligned} L_j(\gamma_j) &= \ell(\gamma_j)\int_0^1 c_j(\gamma_j(t))^{1/2}\, dt \\ &\ge \ell(\gamma_j)\int_0^1 c_i(\gamma_j(t))^{1/2} dt, \end{aligned}$$

and therefore,

$$\begin{aligned} \lim_{j\to\infty} \rho_{c_j}(x,y) &\ge \liminf_{j\to\infty} L_j(\gamma_j) \\ &\ge \liminf_{j\to\infty} \ell(\gamma_j)\int_0^1 c_i(\gamma_j(t))^{1/2} dt \\ &= \ell \int_0^1 \lim_{j\to\infty} c_i(\gamma_j(t))^{1/2} dt \\ &= \ell \int_0^1 c_i(\gamma(t))^{1/2} dt \\ &\ge \int_0^1 c_i(\gamma(t))^{1/2} |\dot{\gamma}(t)|\, dt, \end{aligned} \tag{4.17}$$

where the first inequality follows from (4.11), the third and fourth inequality from the uniform convergence of the functions $\gamma_j(t)$ on $[0,1]$ and the last inequality follows from (4.16). Finally we let $i \to \infty$ and apply the monotone convergence theorem. Denoting by $L(\gamma)$ the length of the path γ in the metric $c(x)ds^2$ it follows from (4.17) that

$$\begin{aligned} \lim_{j\to\infty} \rho_{c_j}(x,y) &\ge \int_0^1 c(\gamma(t))^{1/2} |\dot{\gamma}(t)|\, dt \\ &= L(\gamma) \\ &\ge \rho_c(x,y) \end{aligned}$$

which completes the proof of the lemma when the $c_j(x)$ are continuous.

We now remove the temporarily added continuity assumption on the $c_j(x)$. Thus suppose that $c_j(x)$, $j = 1,2,\cdots$, is a locally bounded lower semicontinuous function such that $c_j(x) \ge \delta > 0$ on $\mathbb{R}^n$. By a well known theorem on semicontinuous functions (see [30; pp. 149-156]) there exists for every fixed j an increasing sequence of continuous functions $c_{jk}(x)$ on $\mathbb{R}^n$, $k = 1,2,\cdots$, such that $c_{jk}(x) \uparrow c_j(x)$ as $k \uparrow \infty$ for all x. Set

$$b_j(x) = \max(c_{1j}(x), c_{2j}(x), \ldots, c_{jj}(x), \delta),$$

$j = 1,2,\cdots$. Then $\{b_j(x)\}$ is an increasing sequence of continuous functions such that

$$0 < \delta \leq b_j(x) \leq c_j(x). \tag{4.18}$$

Also $b_j(x) \geq c_{ij}(x)$ for $i \leq j$. Hence fixing i we have

$$\lim_{j \to \infty} b_j(x) \geq \lim_{j \to \infty} c_{ij}(x) = c_i(x)$$

which in view of (4.18), since i is arbitrary, implies that

$$\lim_{j \to \infty} b_j(x) = \lim_{j \to \infty} c_j(x) = c(x). \tag{4.19}$$

We are now in a position to apply the special case of the lemma already proved to the sequence of distance functions ρ_{b_j}. It follows from (4.19) that

$$\lim_{j \to \infty} \rho_{b_j}(x,y) = \rho_c(x,y)$$

for every $(x,y) \in \mathbb{R}^n \times \mathbb{R}^n$. Since $\rho_{b_j}(x,y) \leq \rho_{c_j}(x,y) \leq \rho_c(x,y)$ this implies

$$\lim_{j \to \infty} \rho_{c_j}(x,y) = \rho_c(x,y)$$

which completes the proof. ▪

We can now prove the following exponential decay theorem in which Lemma 2.8 is used to provide a non constant λ function.

4.4 Theorem: *Let* $P = -\sum_{i,j=1}^{n} a^{ij}\partial_i\partial_j + q(x)$ *be an elliptic operator on* $\mathbb{R}^n$ *such that* $[a^{ij}]$ *is a positive definite constant matrix,* q *a real function in* $L^1_{loc}(\mathbb{R}^n)$ *with* $q_- \in M_{loc}(\mathbb{R}^n)$. *Suppose that* $\Sigma(P) > -\infty$. *Let* $\Omega = \mathbb{R}^n \setminus K$ *where* K *is some compact set and let* $z \in \mathbb{C}$ *with* $\operatorname{Re} z = \mu < \Sigma(P)$. *Suppose* $P\psi = z\psi$ *in* Ω *in the sense that* $\psi \in H^1_{loc}(\Omega)$, $q\psi \in L^1_{loc}(\Omega)$ *and*

$$\int_\Omega (\nabla_A\psi \cdot \nabla_A\varphi + q\psi\varphi)dx = \int_\Omega z\psi\varphi dx$$

for every $\varphi \in C_0^\infty(\Omega)$. *Fix* $N > \Sigma(P) - \mu$ *and set*

$$c(x) = \begin{cases} \min(K(x/|x|)-\mu, N) & \text{if } x \neq 0 \\ 0 & \text{if } x = 0 \end{cases}$$

where $K(\omega) = K(\omega,P)$ *is defined by* (2 21) *Let* $\rho_c(x)$ *denote the distance in* $\mathbf{R}^n$ *from* 0 *to* x *in the metric* $ds_c^2 = c(x)\sum_{i,j} a_{ij}\,dx_i dx_j$ *where* $[a_{ij}] = [a^{ij}]^{-1}$ *(Note that* $c(x) \geq \Sigma(P)-\mu > 0$ *for all* $x \neq 0$ *in view of Lemma 2 7) Suppose*

$$\int_\Omega |\psi(x)|^2 e^{-2(1-\delta)\rho_c(x)}dx < \infty \tag{4 20}$$

for some $0 < \delta < 1$ *Then*

$$\int_\Omega |\psi(x)|^2 e^{2(1-\varepsilon)\rho_c(x)}dx < \infty \tag{4 21}$$

for any $\varepsilon > 0$

Remark. The function $K(\omega) = K(\omega,P)$ was initially defined for all unit-vectors ω in the inner product space $\mathbf{R}^n$ From their definition it is clear, however, that the functions $K(x) = K(x/|x|)$ and $c(x)$ are independent of the special inner product imposed on $\mathbf{R}^n$ This trivial observation should be kept in mind throughout this chapter

Proof of Theorem 4 4 The first step in the proof is to approximate ρ_c by some ρ_{c_j} where c_j is continuous on $\mathbf{R}^n \setminus \{0\}$ Let $\{c_j(\omega)\}$ be a non-decreasing sequence of continuous functions on $S^{n-1} \subset \mathbf{R}^n$ such that $0 < c_j(\omega) < c(\omega)$ and $\lim_{j\to\infty} c_j(\omega) = c(\omega)$ for all $\omega \in S^{n-1}$ Such a sequence exists by the lower semicontinuity of $c(\omega)$ We also denote by c_j the extensions of $c_j(\omega)$ to $\mathbf{R}^n$ defined by

$$c_j(x) = \begin{cases} c_j(x/|x|) & \text{if } x \neq 0 \\ 0 & \text{if } x = 0 \end{cases}$$

The sequence $c_j(x)$ satisfies the hypotheses of Lemma 4 3 Therefore

$$\lim_{j\to\infty} \rho_{c_j}(x) = \rho_c(x)$$

uniformly on compact subsets of $\mathbf{R}^n$ where $\rho_{c_j}(x)$ denotes the distance from 0 to x in the metric $ds_{c_j}^2$ Therefore, given $0 < \varepsilon < 1$ there exists j_0 such that

$$(1-\frac{\varepsilon}{2})\rho_c(\omega) \le \rho_{c_{j_0}}(\omega) \le \rho_c(\omega)$$

for $\omega \in S^{n-1} \subset \mathbb{R}^n$. It follows from their definition that $\rho_{c_j}(x)$ are homogeneous functions on $\mathbb{R}^n$. Therefore

$$(1-\frac{\varepsilon}{2})\rho_c(x) \le \rho_{c_{j_0}}(x) \le \rho_c(x) \tag{4.22}$$

for every $x \in \mathbb{R}^n$.

Next observe that since $c_{j_0}(\omega) < c(\omega) \le K(\omega;P) - \mu$, we have: $c_{j_0}(\omega) + \mu < K(\omega;P)$. Therefore, by Lemma 2.8, there exists $R_0 > 0$ such that

$$\int_{\Omega_0}(|\nabla_A\varphi|^2 + (q-\mu)|\varphi|^2)dx \ge \int_{\Omega_0} c_{j_0}(x)|\varphi|^2 dx \tag{4.23}$$

for all $\varphi \in C_0^\infty(\Omega_0)$ where $\Omega_0 = \{x : |x| > R_0\}$. Choosing R_0 large enough we can also assume that $\overline{\Omega}_0 \subset \Omega$. We are now in a position to apply Theorem 1.5 to the function ψ in Ω_0. Indeed, the inequality (4.23) shows that the hypothesis for $\lambda(x)$ in Theorem 1.5 holds with $\lambda(x) = c_{j_0}(x)$. The function $q(x)-\mu$ satisfies the hypothesis for $q(x)$ in Theorem 1.5 and ψ is a solution of $A\psi + (q-\mu)\psi = 0$ in Ω_0. We set $h(x) = \vartheta\rho_{c_{j_0}}(x)$ where $0 < \vartheta < 1$ then, using Theorem 1.4, we have a.e.

$$|\nabla_A h(x)|^2 = \vartheta^2|\nabla_A\rho_{c_{j_0}}(x)|^2 \le \vartheta^2 c_{j_0}(x) < c_{j_0}(x)$$

which shows that $\vartheta\rho_{c_{j_0}}(x)$ satisfies the hypothesis for h in Theorem 1.5. It remains to show that for some $0 < \delta' < 1$

$$\int_{\Omega_0}|\nabla(\psi)|^2 c_{j_0}(x)\exp(-2(1-\delta')\tilde{\rho}_{c_{j_0}}(x))dx < \infty \tag{4.24}$$

where $\tilde{\rho}_{c_{j_0}}(x) = \tilde{\rho}_{c_{j_0}}(x,x^0)$ denotes the distance in Ω_0 from some $x^0 \in \Omega_0$ to x in the metric $ds^2_{c_{j_0}}$ restricted to Ω_0. Since $\rho_{c_{j_0}}$ denotes the distance function in $\mathbb{R}^n$ with the same metric, we have

$$\begin{aligned}\tilde{\rho}_{c_{j_0}}(x) = \tilde{\rho}_{c_{j_0}}(x,x^0) &\ge \rho_{c_{j_0}}(x,x^0)\\ &\ge \rho_{c_{j_0}}(x,0) - \rho_{c_{j_0}}(x^0,0)\end{aligned}$$

$$\geq (1-\frac{\varepsilon}{2})\rho_c(x) - C$$

where we have used (4.22) to derive the last inequality and where C does not depend on x. Thus

$$2(1-\delta')\tilde{\rho}_{c_{j_0}}(x) \geq 2(1-\delta')(1-(\varepsilon/2))\rho_c(x) - C'$$

where C' does not depend on x. We choose δ' so that

$$(1-\delta')(1-(\varepsilon/2)) = 1-\delta$$

i.e.,

$$\delta' = (\delta-(\varepsilon/2)) / (1-(\varepsilon/2)).$$

We can assume without loss of generality that ε is small enough so that $0 < \delta' < 1$. Then $2(1-\delta')\tilde{\rho}_{c_{j_0}}(x) \geq 2(1-\delta)\rho_c(x) - C'$ so (4.24) follows from the hypothesis (4.20).

Therefore we can apply Theorem 1.5 to conclude

$$\int_{\Omega_{0,d}} |\psi(x)|^2(c_{j_0}(x)-|\nabla_A h(x)|^2)e^{2h(x)}dx$$

$$\leq \frac{2(1+2d)}{d^2} \int_{\Omega_0 \setminus \Omega_{0,d}} |\psi(x)|^2 c_{j_0}(x) e^{2h(x)}dx$$

for $d > 0$ and $\Omega_{0,d} = \{x \in \Omega_0\colon \tilde{\rho}_{c_{j_0}}(x,\partial\Omega_0) > d\}$.

Now for $x \in \Omega$

$$c_{j_0}(x)-|\nabla_A h(x)|^2 \geq (1-\vartheta^2)c_{j_0}(x)$$

$$\geq (1-\vartheta^2) \inf_{|\omega|=1} c_{j_0}(\omega) = m$$

where m is a positive constant. Also,

$$2h(x) = 2\vartheta\rho_{c_{j_0}}(x) \geq 2\vartheta(1-\frac{\varepsilon}{2})\rho_c(x),$$

so that if we now choose $\vartheta = (1-\varepsilon)/(1-\frac{\varepsilon}{2})$ then $0 < \vartheta < 1$ and

$$2h(x) \geq 2(1-\varepsilon)\rho_c(x).$$

Therefore

$$\int_{\Omega_{0,d}} |\psi(x)|^2 e^{2(1-\varepsilon)\rho_c(x)} dx$$

$$\leq m^{-1} \int_{\Omega_{0,d}} |\psi(x)|^2 (c_{j_0}(x) - |\nabla_A h(x)|^2) e^{2h(x)} dx$$

$$\leq \frac{2(1+2d)}{md^2} \int_{\Omega_0 \setminus \Omega_{0,d}} |\psi(x)|^2 c_{j_0}(x) e^{2h(x)} dx. \tag{4.25}$$

Since $\Omega_0 \setminus \Omega_{0,d}$ is contained in a compact subset of Ω the condition $\psi \in L^2_{loc}(\Omega)$ implies the right side of (4.25) is finite. We also have

$$\int_{\Omega \setminus \Omega_{0,d}} |\psi(x)|^2 e^{2(1-\varepsilon)\rho_c(x)} dx$$

$$\leq C_2 \int_{\Omega \setminus \Omega_{0,d}} |\psi(x)|^2 e^{-2(1-\delta)\rho_c(x)} dx$$

$$< \infty,$$

where the first inequality follows from the continuity of $\rho_c(x)$ on $\mathbb{R}^n$ and the fact that $\Omega \setminus \Omega_{0,d}$ is contained in a compact subset of $\mathbb{R}^n$ and the second inequality follows from the hypothesis (4.20). Therefore

$$\int_{\Omega} |\psi(x)|^2 e^{2(1-\varepsilon)\rho_c(x)} dx < \infty$$

which completes the proof. ▪

Theorem 4.4 has the following corollary for L^2 eigenfunctions.

4.5 Corollary: *Let* $P = -\sum_{i,j=1}^{n} a^{ij} \partial_i \partial_j + q(x)$ *be an elliptic operator on* $\mathbb{R}^n$ *satisfying the hypotheses of Theorem 4.4. Suppose also that* $K(\omega) = K(\omega; P)$ *is a bounded function on* S^{n-1}. *Let* H *be the self-adjoint realization of* P *in* $L^2(\mathbb{R}^n)$ *defined in Theorem 3.2. Let* ψ *be an eigenfunction of* H *with eigenvalue* $\mu < \inf \sigma_{ess}(H)$. *Then*

$$\int_{\mathbb{R}^n} |\psi(x)|^2 e^{2(1-\varepsilon)\rho_c(x)} dx < \infty \tag{4.26}$$

for any $\varepsilon > 0$ *where* $c(x) = K(x/|x|) - \mu$ *and* $\rho_c(x)$ *is defined as in Theorem 4.4.*

We now want to apply these results to eigenfunctions of N-body quantum systems. Since the previous theorems gave exponential decay in terms of $K(\omega;P)$, our goal will be to describe this function when P satisfies certain properties which hold for quantum mechanical N-body Hamiltonians. We also change notation to conform with conventions in physics: $q(x)$ is rechristened $V(x)$ and is called the potential and we let $\Delta = \sum_{i,j=1}^{n} a^{ij}\partial_i\partial_i$. Then $P = -\Delta + V(x)$. Note that Δ is the Laplace-Beltrami operator associated with the inner product given on $\mathbb{R}^n$ by $x \cdot y = \sum_{i,j=1}^{n} a_{ij}x_i y_i$, where as usual $[a_{ij}] = [a^{ij}]^{-1}$.

4.6 Lemma: *Let $\Lambda(P)$, $\Sigma(P)$, and $K(\omega) = K(\omega;P)$ be as defined in Definition 2.4 and Definition 2.6. Suppose that $V(x+t\omega_0) = V(x)$ for some $\omega_0 \in S^{n-1} \subset \mathbb{R}^n$ and all $t \in \mathbb{R}$. Then*

(i) $\Lambda(P) = \Sigma(P)$

(ii) $K(\omega_0;P) = \Sigma(P) = \min\{K(\omega) : \omega \in S^{n-1}\}$.

Proof: Let $\varphi \in C_0^\infty(\mathbb{R}^n)$ and define $\varphi_\tau(x) = \varphi(x+\tau\omega_0)$ for $\tau \in \mathbb{R}$. Then

$$\frac{(P\varphi,\varphi)}{||\varphi||^2} = \frac{(P\varphi_\tau,\varphi_\tau)}{||\varphi_\tau||^2}.$$

To prove (i) we need to show $\Sigma(P) \leq \Lambda(P)$, since the opposite inequality follows from the defintions. Fix $\varepsilon > 0$ and let K be a compact set such that

$$\inf\{\frac{(P\varphi,\varphi)}{||\varphi||^2} : \varphi \in C_0^\infty(\mathbb{R}^n \setminus K)\} > \Sigma - \varepsilon.$$

Pick $\varphi \in C_0^\infty(\mathbb{R}^n)$ such that

$$\frac{(P\varphi,\varphi)}{||\varphi||^2} < \Lambda(P) + \varepsilon.$$

Since φ has compact support there exists τ so that $\varphi_\tau \in C_0^\infty(\mathbb{R}^n \setminus K)$. Thus

$$\Lambda(P) + \varepsilon > \frac{(P\varphi,\varphi)}{||\varphi||^2}$$
$$= \frac{(P\varphi_\tau,\varphi_\tau)}{||\varphi_\tau||^2}$$

$$\geq \inf \{ \frac{(P\varphi,\varphi)}{\|\varphi\|^2} : \varphi \in C_0^\infty(\mathbf{R}^n \setminus K) \}$$

$$> \Sigma - \varepsilon$$

which proves (i) since $\varepsilon > 0$ was arbitrary.

Noting that for any $\varphi \in C_0^\infty(\mathbf{R}^n)$, φ_τ is supported in the truncated cone $\Gamma_{\omega_0}^{\varepsilon,N}$ (see Definition (2.6)) for any ε and N when τ is large enough, the first equality in (ii) follows from a similar argument. The second equality is equation (2.22) so the proof is complete. ▪

In the N-body problem the potential $V(x)$ on $\mathbf{R}^n$ is of the form $V(x) = \sum_{i=1}^{\ell} V_i(x)$ where each of the potentials $V_i(x)$ depends only on some of the variables. We shall need the following lemma to show that if V_i belongs to the class M_δ on the subspace on which it depends then $V_i \in M_\delta(\mathbf{R}^n)$. For the definitions of the classes of functions $M_\delta(\Omega)$ for $0 \leq \delta < 1$ we refer to Chapter 0. We also recall that $M_0(\Omega)$ is another notation for the class of functions $M(\Omega)$.

4.7 Lemma: *Let $\mathbf{R}^n = Y \oplus Z$ where Y and Z are two complementary subspaces of $\mathbf{R}^n$. For a point $x \in \mathbf{R}^n$ we use the obvious notation $x = (y, z)$. Suppose V is a function on $\mathbf{R}^n$ depending only on y*, i.e.,

$$V((y,z)) = W(y)$$

and $W \in M_\delta(Y)$ for some $0 \leq \delta < 1$. Then $V \in M_\delta(\mathbf{R}^n)$.

Proof: We consider $\mathbf{R}^n$ as an inner product space with norm $|x|$ and associated measure dx. We denote by dy and dz the measures induced by dx on Y and Z respectively. We have $dx = \alpha dy \otimes dz$ where α is some positive constant. Let $m = \dim Y$. We give the proof under the assumption that $2 - \delta < m < n$. The necessary modifications needed in the proof for $m = 1$ and $m = 2$ are left to the reader.

We shall estimate the integral appearing in the definition of $M_\delta(\mathbb{R}^n)$. (We note here that the definition of the class of functions $M_\delta(\mathbb{R}^n)$ is independent of the inner product imposed on $\mathbb{R}^n$). For any $x^0 = (y^0, z^0) \in \mathbb{R}^n$ and $r > 0$, we have

$$\int_{|x-x^0|<r} |V(x)|\,|x-x^0|^{2-n-\delta}dx \leq$$

$$\alpha \int_{|y-y^0|<r} |W(y)| \int_Z (|y-y^0|^2 + |z-z^0|^2)^{(2-n-\delta)/2} dz dy$$

$$= \alpha c \int_{|y-y^0|<r} |W(y)|\,|y-y^0|^{2-m-\delta}dy$$

where c is a constant depending only on $n+\delta$ and $n-m$. Since $W \in M_\delta(Y)$ the right side is finite and tends to 0 as $r \to 0$ uniformly in y^0. Therefore the left side tends to 0 as $r \to 0$ uniformly in x^0. Thus $V \in M_\delta(\mathbb{R}^n)$. ▪

4.8 Lemma: *Suppose* Π_i $i = 1, \ldots, \ell$ *are non-zero projections on* $\mathbb{R}^n$ *such that* $\Pi_i \neq \Pi_j$ *if* $i \neq j$. *Suppose* $V_i(x)$ $i = 1, \ldots, \ell$ *are real functions on* $\mathbb{R}^n$ *such that*

(i) $V_i(x) = V_i(\Pi_i x)$.

(ii) $V_i(x) \to 0$ as $|\Pi_i x| \to \infty$.

Set $Y_i = \mathrm{Ran}\Pi_i$ *and suppose that* $V_i \restriction Y_i \in L^1_{loc}(Y_i)$ *and* $(V_i)_- \restriction Y_i \in M_{loc}(Y_i)$. *Set* $V = \sum_{i=1}^{\ell} V_i$ *and let* $P = -\Delta + V = -\sum_{i,j=1}^{n} a^{ij}\partial_i\partial_j + V(x)$. *Then for any* $\omega_0 \in S^{n-1} \subset \mathbb{R}^n$:

$$K(\omega_0;P) = K(\omega_0;P_{\omega_0}) = \Sigma(P_{\omega_0}) = \Lambda(P_{\omega_0}) \tag{4.27}$$

where

$$P_{\omega_0} = -\Delta + \sum_{\Pi_i\omega_0 = 0} V_i.$$

(Here and in the following $\sum_{\Pi_i\omega=0} V_i$ *denotes summation of* V_i *over those indices* i *for which* $\Pi_i\omega = 0$.)

Before proving the Lemma we draw some conclusions from (4.27). First observe that $K(\omega;P)$ takes only a finite number of values as ω ranges over the unit sphere. This follows since there are only a finite number of operators P_ω as ω ranges over S^{n-1} in $\mathbb{R}^n$. Next, let

$$E = \{\omega \in S^{n-1} : \Pi_i \omega \neq 0 \quad \text{for} \quad i = 1, \ldots, \ell\}.$$

By definition we have

$$P_\omega = -\Delta \ \text{for} \quad \omega \in E$$

so that (4.27) gives

$$K(\omega;P) = \Lambda(-\Delta) = 0 \quad \text{for} \ \omega \in E.$$

We note that E is an open dense set in S^{n-1} where the last property follows from the inclusion relation

$$S^{n-1} \setminus E \subset \bigcup_{i=1}^{\ell} (S^{n-1} \cap \ker \Pi_i).$$

Since $K(\omega;P)$ is a lower semicontinuous function on S^{n-1} which vanishes on the dense set E it follows that

$$\max\{K(\omega;P) : \omega \in S^{n-1}\} = 0. \tag{4.28}$$

We also remark that under the conditions of Lemma 4.8 $\Lambda(P) > -\infty$. For by Condition (ii) $(V_i)_- \restriction Y_i$ is the sum of a bounded function and a function of compact support in Y_i. Therefore $(V_i)_- \restriction Y_i \in M(Y_i)$ so that by Lemma 4.7 $\sum_{i=1}^{\ell} (V_i)_- \in M(\mathbb{R}^n)$ which implies, since $(\sum_{i=1}^{\ell} V_i)_- \leq \sum_{i=1}^{\ell} (V_i)_-$, that $V_- \in M(\mathbb{R}^n)$. Applying Lemma 0.3 to V_- it follows that there exists a constant c such that

$$(V_- \varphi, \varphi) \leq (-\Delta \varphi, \varphi) + c\,||\varphi||^2$$

for all $\varphi \in C_0^\infty(\mathbb{R}^n)$. This implies that

$$(P\varphi, \varphi) \geq -c\,||\varphi||^2$$

for all $\varphi \in C_0^\infty(\mathbb{R}^n)$, which gives

$$\Sigma(P) \geq \Lambda(P) > -\infty. \tag{4.29}$$

The same proof shows that $\Lambda(P_1) > -\infty$ for $P_1 = -\Delta + \Sigma' V_i$ where the sum is extended over any subset of $\{1, \ldots, \ell\}$

Proof of Lemma 4.8: Let $\omega_0 \in S^{n-1} \subset \mathbb{R}^n$ be given We want to show that if $\Pi_i \omega_0 \neq 0$ then $V_i \to 0$ uniformly in some cone around ω_0 If this is true then it is clear from the definition of $K(\omega_0,P)$ that V_i can be deleted from the sum $V = \sum_{i=1}^{\ell} V_i$ when calculating $K(\omega_0,P)$ So suppose $0 < \delta < |\Pi_i \omega_0|$ By the continuity of Π_i there is a ball B in S^{n-1} about ω_0 such that if $\omega \in B$, $|\Pi_i \omega| \geq |\Pi_i \omega_0| - \delta$ Condition (ii) means that for every $\varepsilon > 0$ there exists $N_\varepsilon > 0$ such that $|\Pi_i x| > N_\varepsilon$ implies $|V_i(x)| < \varepsilon$ If $L \geq N_\varepsilon / (|\Pi_i \omega_0| - \delta)$ and C is the cone defined by

$$C = \{t\omega \quad t > 0 \, , \, \omega \in B\},$$

then if $x \in C$ and $|x| > L$, we have

$$\begin{aligned} |\Pi_i x| &= |x| \, |\Pi_i (\frac{x}{|x|})| \geq |x| (|\Pi_i \omega_0| - \delta) \\ &\geq L(|\Pi_i \omega_0| - \delta) \geq N_\varepsilon \end{aligned}$$

so that $|V_i(x)| < \varepsilon$ Therefore $\lim_{\substack{|x| \to \infty \\ x \in C}} V_i(x) = 0$ So if we set

$$P_{\omega_0} = -\Delta + \sum_{\Pi_i \omega_0 = 0} V_i$$

we have $K(\omega_0,P) = K(\omega_0,P_{\omega_0})$ But if $\Pi_i \omega_0 = 0$

$$\begin{aligned} V_i(x+t\omega_0) &= V_i(\Pi_i(x+t\omega_0)) \\ &= V_i(\Pi_i x) \\ &= V_i(x), \end{aligned}$$

so we can apply Lemma 4 6 to conclude

$$K(\omega_0,P) = K(\omega_0,P_{\omega_0}) = \Sigma(P_{\omega_0}) = \Lambda(P_{\omega_0}),$$

which completes the proof ▪

We shall refer to the operator $P = -\Delta + V$ introduced in Lemma 4 8 as a multiparticle type Schrodinger operator Under the assumptions of the Lemma $V \in L^1_{loc}(\mathbb{R}^n)$ and $V_- \in M_{loc}(\mathbb{R}^n)$ We also have $\Lambda(P) > -\infty$ Thus Theorem 3 2 is applicable to P We shall denote by H the self-adjoint realization of P in $L^2(\mathbb{R}^n)$ defined in Theorem 3.2. H is bounded from below and is unique in some general sense. We refer to H as *the* self-adjoint realization of P. Similarly, since $\Sigma(P_\omega) = \Lambda(P_\omega) > -\infty$, Theorem 3.2 is applicable to the operators P_ω. We denote by H_ω the self-adjoint realization of P_ω in $L^2(\mathbb{R}^n)$. From Lemma 4.8 and Theorem 3.2 it follows that

$$K(\omega;P) = \Sigma(P_\omega) = \inf \sigma(H_\omega) \tag{4.30}$$

for every $\omega \in S^{n-1}$.

Applying Corollary 4.5 and using (4.30) (note that $\Sigma(P) \leq K(\omega;P) \leq 0$ in view of (4.28) and (2.22)) we obtain the following theorem on exponential decay of eigenfunctions of multiparticle type Schrödinger operators.

4.9 Theorem: *Let $P = -\Delta + V$ be a multiparticle type Schrödinger operator*

on $\mathbb{R}^n$. $\Delta = -\sum_{i,j} a^{ij}\partial_i\partial_j$, $V(x) = \sum_{i=1}^{\ell} V_i(x)$ *where the $V_i(x)$ satisfy the hypotheses of Lemma* 4.8. *Let H be the self-adjoint realization of P in $L^2(\mathbb{R}^n)$ and let H_ω be the self-adjoint realization of $P_\omega = -\Delta + \sum_{\Pi_i\omega=0} V_i$, $\omega \in S^{n-1} \subset \mathbb{R}^n$. Set*

$$\Sigma_\omega = \inf \sigma(H_\omega).$$

Let ψ be an eigenfunction of H with eigenvalue $\mu < \inf \sigma_{ess}(H)$. Then

$$\int_{\mathbb{R}^n} |\psi(x)|^2 \, e^{2(1-\varepsilon)\rho(x)} dx < \infty \tag{4.31}$$

for any $\varepsilon > 0$ where $\rho(x)$ is the geodesic distance from x to 0 in the Riemannian metric

$$ds^2 = c(x) \sum_{i,j=1}^{n} a_{ij}\, dx_i\, dx_j \tag{4.32}$$

with $c(x) = \Sigma_{x/|x|} - \mu$ and $[a_{ij}] = [a^{ij}]^{-1}$.

Remark: The function $c(x)$ which appears in the definition of the Riemannian metric is defined on $\mathbb{R}^n \setminus \{0\}$ where it is positive bounded and lower semicontinuous. We have neglected to define $c(x)$ for $x = 0$ and to be exact we should formally set $c(0) =$ some non-negative number. From remarks made before it is clear that $\rho(x)$ does not depend on the special value chosen for $c(0)$.

We shall now consider some special multiparticle Schrödinger operators arising in physical problems.

We consider an atomic type system of $N+1$ particles with coordinates $x^i \in \mathbf{R}^\nu$, $i = 0,1, \ldots, N$, and interacting potentials $v_{ij}(y)$ defined on $\mathbf{R}^\nu$. The particle x^0 (the "nucleous") has infinite mass and is fixed at x^0. We assume that the real valued functions v_{ij}, $0 \le i < j \le N$, verify the following conditions:

(i) $v_{ij} \in L^1_{loc}(\mathbf{R}^\nu)$ and $\lim_{|y| \to \infty} v_{ij}(y) = 0$ for $0 \le i < j \le N$.

(ii) $v_{ij} \ge 0$ for $1 \le i < j \le N$.

(iii) $(v_{01})_- \in M_{loc}(\mathbf{R}^\nu)$ for $1 \le i \le N$.

The configuration space of the system consists of the product of N copies of $\mathbf{R}^\nu$ identified with $\mathbf{R}^{\nu N}$ which we also denote by X. The generic point in X is $x = (x^1, \ldots, x^N)$ where $x^i = (x^i_1, \ldots, x^i_\nu) \in \mathbf{R}^\nu$ are the coordinates of the particles. The Schrödinger operator of the system is the elliptic operator P on X defined by

$$P = -\sum_{i=1}^{N} (2m_i)^{-1} \Delta_i + \sum_{i=1}^{N} v_{0i}(x^i) + \sum_{1 \le i < j \le N} v_{ij}(x^i - x^j) \tag{4.33}$$

where Δ_i is the ordinary Laplace operator in x^i and the m_i are positive numbers representing the masses of the particles. The potential of the system is the function V on X given by

$$V(x) = \sum_{0 \le i < j \le N} V_{ij}(x)$$

where

$$V_{0j}(x) = v_{0j}(x^j) \quad \text{for} \quad 1 \le j \le N$$
$$V_{ij}(x) = v_{ij}(x^i - x^j) \quad \text{for} \quad 1 \le i < j \le N.$$

We define in X a new inner product For $x = (x^1, \ldots, x^N)$, $y = (y^1, \ldots, y^N)$

$$<x,y> = \sum_{i=1}^{N} 2m_i x^i \, y^i \tag{4 34}$$

where $x^i y^i$ denotes the usual inner product in $\mathbf{R}^\nu$ We denote by Δ the Laplace-Beltrami operator in X with respect to the inner product (4 34) With this notation the Schrodinger operator (4 33) can be written in the form

$$P = -\Delta + V = -\sum_{i=1}^{N} (2m_i)^{-1} \Delta_i + \sum_{0 \le i < j \le N} V_{ij}$$

We introduce in X projection operators Π_{ij}, $0 \le i < j \le N$, defined as follows

$$(\Pi_{0j} x)^k = \begin{cases} x^k & \text{if } k = j \\ 0 & \text{if } k \ne j \end{cases}$$

for $j = 1, \ldots, N$, and

$$(\Pi_{ij} x)^k = \begin{cases} m_j (x^i - x^j)/(m_i + m_j) & \text{if } k = i \\ m_i (x^j - x^i)/(m_i + m_j) & \text{if } k = j \\ 0 & \text{if } k \ne i \text{ or } j \end{cases} \tag{4 35}$$

for $1 \le i < j \le N$ The reader can check that Π_{ij} is an orthogonal projection in the inner product (4 34) In addition we have

$$V_{ij}(\Pi_{ij} x) = V_{ij}(x) \tag{4 36}$$

for $0 \le i < j \le N$

The above considerations show that Lemma 4 8 is applicable to the atomic type Schrodinger operator introduced in (4 33) Applying the lemma we conclude that for any unit-vector $\omega \in X$

$$K(\omega, P) = \Lambda(P_\omega) \tag{4 37}$$

where P_ω is the Schrodinger operator (on X) defined by

$$P_\omega = -\Delta + \sum_{\substack{0 \le i < j \le N \\ \Pi_{ij}\omega = 0}} V_{ij} \tag{4 38}$$

Using (4 37) we could now apply Theorem 4 9 to eigenfunctions of P However, we prefer to obtain first some additional information on the structure of the function $K(\omega,P)$ and then apply Corollary 4 5 directly to derive the non-isotropic exponential bounds for eigenfunctions of P

We shall use the following notation Let $x = (x^1, \ldots, x^N) \in X$ Then $I(x)$ will denote the subset of integers i in the set $\{1, \ldots, N\}$ such that $x^i = 0$ ($I(x)$ is the empty set if $x^i \neq 0$ for all i) Given a unit-vector $\omega = (\omega^1, \ldots, \omega^N)$, we denote by $\widetilde{P}_\omega$ the Schrodinger operator on X defined by

$$\widetilde{P}_\omega = -\Delta + \sum_{i \in I(\omega)} V_{0i} + \sum_{\substack{1 \le i < j \le N \\ i,j \in I(\omega)}} V_{ij} \tag{4 39}$$

Noting that

$$\sum_{\substack{0 \le i < j \le N \\ \Pi_{ij}\omega = 0}} V_{ij} = \sum_{\substack{1 \le j \le N \\ \omega^j = 0}} V_{0j} + \sum_{\substack{1 \le i < j \le N \\ \omega^i = \omega^j}} V_{ij},$$

it follows from (4 38) and (4 39) that

$$P_\omega = \widetilde{P}_\omega + \widetilde{V}_\omega \tag{4 40}$$

where

$$\widetilde{V}_\omega = \sum_{\substack{1 \le i < j \le N \\ \omega^i = \omega^j \\ \omega^i \neq 0}} V_{ij}$$

We shall first show that

$$\Lambda(P_\omega) = \Lambda(\widetilde{P}_\omega) \tag{4 41}$$

for every unit-vector ω in X

Now, by assumption $V_{ij} \geq 0$ for $i,j \geq 1$ which implies that $\widetilde{V}_\omega \geq 0$ and thus, in view of (4 40) and (2 11), it follows that

$$\Lambda(P_\omega) \geq \Lambda(\widetilde{P}_\omega) \tag{4 42}$$

To establish the reverse inequality we fix a unit-vector $\omega = (\omega^1, \omega^2, \ldots, \omega^N)$ in X. Set $\omega_* = (\omega^1, 2\omega^2, \ldots, N\omega^N)$ and introduce a sequence of unit-vectors $\omega_k = (\omega_k^1, \ldots, \omega_k^N)$ in X defined by

$$\omega_k = (\omega + k^{-1}\omega_*)/|\omega + k^{-1}\omega_*| \quad \text{for } k = 1,2,$$

It is clear that $\omega_k \to \omega$ as $k \to \infty$ and that

(a) $I(\omega_k) = I(\omega)$ for all $k \geq 1$.

It is also easy to see that there exists an integer $k_0 \geq 1$ such that

(b) $\omega_k^i \neq \omega_k^j$ for $1 \leq i < j \leq N$ and $k \geq k_0$ except when both i and j belong to $I(\omega)$ (in which case $\omega_k^i = \omega_k^j = 0$).

From (a) and (b) it follows, in view of (4.39) and (4.40), that

$$P_{\omega_k} = \widetilde{P}_{\omega_k} = \widetilde{P}_\omega \tag{4.43}$$

for $k \geq k_0$. Applying (4.37) (with $\omega = \omega_k$) and using (4.43) it follows that

$$K(\omega_k;P) = \Lambda(P_{\omega_k}) = \Lambda(\widetilde{P}_\omega) \tag{4.44}$$

for $k \geq k_0$. Letting $k \to \infty$ it follows from (4.44) and the lower semicontinuity of the function K (since $\omega_k \to \omega$), using (4.37), that

$$\Lambda(\widetilde{P}_\omega) = \lim_{\omega_k \to \omega} K(\omega_k;P) \geq K(\omega;P) = \Lambda(P_\omega),$$

which when combined with (4.42) proves (4.41).

Next we rewrite the operator $\widetilde{P}_\omega$ in the following form:

$$\widetilde{P}_\omega = -\widetilde{\Delta}_\omega + Q_\omega \tag{4.45}$$

where

$$\widetilde{\Delta}_\omega = \sum_{\substack{1 \leq i \leq N \\ i \notin I(\omega)}} (2m_i)^{-1} \Delta_i ,$$

$$Q_\omega = \sum_{i \in I(\omega)} (-(2m_i)^{-1}\Delta_i + v_{0i}(x^i)) + \sum_{\substack{i<j \\ i,j \in I(\omega)}} v_{ij}(x^i - x^j) \tag{4.46}$$

if $I(\omega)$ is not empty, $Q_\omega = 0$ if $I(\omega)$ is empty. We also associate with every unit-vector ω a subspace $X_\omega \subset X$ defined by

$$X_\omega = \{x = (x^1, \ldots, x^N) \in X : x^i = 0 \text{ if } i \notin I(\omega), 1 \leq i \leq N\}. \tag{4.47}$$

From its definition it is clear that the operator Q_ω is translation invariant in directions orthogonal to X_ω and that it has a natural restriction to X_ω. In the following we denote by Q_ω' the restriction of Q_ω to X_ω. The operator Q_ω' can also be described as the Schrödinger operator of a subsystem of particles with a configuration space X_ω. The following lemma shows that the bottom of the spectrum of $Q_\omega^{'}$ is the same as the bottom of the spectrum of $\widetilde{P}_\omega$

4.10 Lemma: *Let $\widetilde{P}_\omega$ and $Q_\omega^{'}$ be defined as above Then*

$$\Lambda(\widetilde{P}_\omega) = \Lambda(Q_\omega^{'}) \tag{4 48}$$

for every unit-vector ω where we set $\Lambda(Q_\omega^{'}) = 0$ if $I(\omega)$ is empty (i e , when $\omega^i \neq 0$ for $i = 1,\ \ \ , N$)

Proof: We recall that X is an inner product space with an inner product given by (4 34) Suppose first that $I(\omega)$ is empty Then it follows from (4 45) that $\widetilde{P}_\omega = -\sum_{i=1}^{N} \frac{1}{2m_i}\Delta_i = -\Delta$ which implies that $\Lambda(\widetilde{P}_\omega) = \Lambda(-\Delta) = 0$ by a well known result This proves (4 48) when $I(\omega)$ is empty

Suppose next that $I(\omega)$ is not empty To simplify notation we denote the subpsace X_ω by Y We denote by Z the orthogonal complement of Y in X and we use the obvious notation $x = (y,z)$ for $x \in X = Y \oplus Z$ We denote by dx , dy and dz the measures induced by the inner product on X, Y and Z, respectively

Let $\varphi(y,z) \in C_0^\infty(X)$ and set $\psi_z(y) = \varphi(y,z)$ so that $\psi_z(y) \in C_0^\infty(Y)$ for every fixed z From the definition of $\Lambda(Q_\omega^{'})$ it follows (recall that $Q_\omega^{'}$ acts on Y)

$$(Q_\omega^{'}\psi_z,\psi_z)_{L^2(Y)} \geq \Lambda(Q_\omega^{'})\,||\psi_z||_{L^2(Y)}^2$$

Integrating the last inequality with respect to z on Z we find that

$$\begin{aligned}(Q_\omega^{'}\varphi,\varphi)_{L^2(X)} &= \int_Z (Q_\omega^{'}\psi_z,\psi_z)_{L^2(Y)}dz \\ &\geq \Lambda(Q_\omega^{'})\int_Z ||\psi_z||_{L^2(Y)}^2 dz \\ &= \Lambda(Q_\omega^{'})\,||\varphi||_{L^2(X)}^2\end{aligned} \tag{4 49}$$

In view of (4 45) we have

$$\begin{aligned}(\widetilde{P}_\omega\varphi,\varphi)_{L^2(X)} &= (-\widetilde{\Delta}_\omega\varphi,\varphi)_{L^2(X)} + (Q_\omega\varphi,\varphi)_{L^2(X)} \\ &\geq (Q_\omega\varphi,\varphi)_{L^2(X)}\end{aligned} \tag{4 50}$$

Hence combining (4.49) and (4.50) we find that

$$\Lambda(\widetilde{P}_\omega) = \inf\{\frac{(\widetilde{P}_\omega\varphi,\varphi)_{L^2(X)}}{||\varphi||_{L^2(X)}} : \varphi \in C_0^\infty(X)\ ,\ \varphi \neq 0\}$$
$$\geq \Lambda(Q_\omega'). \tag{4.51}$$

To prove the reverse inequality we fix a number $\delta > 0$ and then choose a function $\chi(y) \in C_0^\infty(Y)$ verifying:

$$||\chi||_{L^2(Y)} = 1,$$
$$(Q_\omega'\chi,\chi)_{L^2(Y)} \leq \Lambda(Q_\omega') + \delta. \tag{4.52}$$

We also pick a function $\zeta(z) \in C_0^\infty(Z)$ such that $||\zeta||_{L^2(Z)} = 1$ and define

$$\zeta_\varepsilon(z) = \varepsilon^{-m/2}\zeta(\varepsilon z) \qquad \text{for} \qquad \varepsilon > 0$$

where $m = \dim Z$. We set:

$$\varphi_\varepsilon(y,z) = \chi(y)\zeta_\varepsilon(z)$$

and note that

$$||\zeta_\varepsilon||_{L^2(Z)} = ||\varphi_\varepsilon||_{L^2(X)} = 1. \tag{4.53}$$

In view of (4.45) we have

$$\widetilde{P}_\omega(\varphi_\varepsilon(y,z)) = \zeta_\varepsilon Q_\omega'\chi - \chi\widetilde{\Delta}_\omega\zeta_\varepsilon. \tag{4.54}$$

A simple computation shows that

$$||\widetilde{\Delta}_\omega\zeta_\varepsilon||_{L^2(Z)} \to 0 \qquad as \qquad \varepsilon \to 0.$$

We now fix a number $\varepsilon_0 > 0$ so that

$$||\widetilde{\Delta}_\omega\zeta_{\varepsilon_0}||_{L^2(Z)} < \delta. \tag{4.55}$$

Using (4.52) through (4.55) it follows that

$$(\widetilde{P}_\omega\varphi_{\varepsilon_0},\varphi_{\varepsilon_0})_{L^2(X)} = (\zeta_{\varepsilon_0}Q_\omega'\chi,\chi\zeta_{\varepsilon_0})_{L^2(X)} - (\chi\widetilde{\Delta}_\omega\zeta_{\varepsilon_0},\chi\zeta_{\varepsilon_0})_{L^2(X)}$$
$$= (Q_\omega'\chi,\chi)_{L^2(Y)} - (\widetilde{\Delta}_\omega\zeta_{\varepsilon_0},\zeta_{\varepsilon_0})_{L^2(Z)}$$
$$\leq \Lambda(Q_\omega') + \delta + ||\widetilde{\Delta}_\omega\zeta_{\varepsilon_0}||_{L^2(Z)}||\zeta_{\varepsilon_0}||_{L^2(Z)}$$
$$< \Lambda(Q_\omega') + 2\delta,$$

which implies (in view of (4.53)) that

$$\Lambda(\tilde{P}_\omega) < \Lambda(Q_\omega') + 2\delta \tag{4 56}$$

Combining (4 56) with (4 51) we obtain (4 48) (since $\delta > 0$ is arbitrary) thus completing the proof ▪

The preceding discussion leads to the following characterization of the function $K(\omega,P)$

4.11 Lemma: *Let P be the atomic type Schrodinger operator defined by* (4 33), *P acts in the configuration space $X = \mathbb{R}^{\nu N}$ For any unit-vector $\omega \in X$ (in some inner product) define the subsystem Schrödinger operator Q_ω' as the restriction of the operator Q_ω (defined by* (4 46)) *to the subspace X_ω (defined by* (4 47)) *Then*

$$K(\omega,P) = \Lambda(Q_\omega') \tag{4 57}$$

Proof. To obtain (4 57) simply combine (4 37) with (4 41) and (4 48) ▪

It was observed before that the atomic type Schrodinger operator P verifies the conditions of Lemma 4 8 Hence it also verifies the conditions of Theorem 3 2 Applying the theorem we shall denote by H the self-adjoint realization of P in $L^2(X)$ If ω is a unit-vector such that $I(\omega)$ is not empty, we shall denote (temporarily) by H_ω' the self-adjoint realization of Q_ω' in $L^2(X_\omega)$ where Q_ω' and X_ω are defined as in Lemma 4 11 It follows from Theorem 3 2 when applied to H_ω' that

$$\Lambda(Q_\omega') = \inf \sigma(H_\omega'),$$

which when combined with (4 57) shows that

$$K(\omega,P) = \inf \sigma(H_\omega') \quad \text{if} \quad I(\omega) \text{ is not empty,} \tag{4 58}$$
$$K(\omega,P) = 0 \quad \text{if} \quad I(\omega) \text{ is empty}$$

We shall now combine Corollary 4 5 with formula (4 58) to derive the exponential decay estimates for eigenfunctions of atomic type Schrodinger operators For reasons of presentation we shall make some changes in the notation used before

4 12 Theorem *Let P be an atomic type Schrodinger operator given by*

$$P = \sum_{i=1}^{N} (-(2m_i)^{-1} \Delta_i + v_{0i}(x^i)) + \sum_{1 \le i < j \le N} v_{ij}(x^i - x^j) \tag{4 59}$$

where $x^i \in \mathbb{R}^\nu$ and Δ_i denotes the ordinary Laplacian in x^i The m_i are positive numbers and the $v_{ij}(y)$ are real valued functions defined on $\mathbb{R}^\nu$ We assume that

(i) $v_{ij} \in L^1_{loc}(\mathbb{R}^\nu)$ *and* $v_{ij}(y) \to 0$ *as* $|y| \to \infty$ *for* $0 \le i < j \le N$

(ii) $v_{ij} \ge 0$ *for* $1 \le i < j \le N$

(iii) $(v_{0i})_- \in M_{loc}(\mathbb{R}^\nu)$ *for* $1 \le i \le N$

We consider P as acting on functions defined on the configuration space X, identified with $\mathbb{R}^{\nu N}$, with generic point $x = (x^1, \quad, x^N)$

Denote by I any proper subset of the set of integers $\{1, \quad, N\}$, I may also be the empty set For every non-empty I denote by X_I the subspace of X defined by

$$X_I = \{x = (x^1, \quad, x^N) \in X \quad x^i = 0 \text{ if } i \notin I\}$$

Denote by P_I the subsystem Schrodinger operator defined on X_I by

$$P_I = \sum_{i \in I} (-(2m_i)^{-1} \Delta_i + v_{0i}(x^i)) + \sum_{\substack{i<j \\ i,j \in I}} v_{ij}(x^i - x^j) \tag{4 60}$$

Let H be the self-adjoint realization of P in $L^2(X)$ and let H_I denote the self-adjoint realization of P_I in $L^2(X_I)$ Set

$$\Lambda_I = \inf \sigma(H_I), \ \Lambda_I = 0 \text{ if } I \text{ is empty} \tag{4 61}$$

For any $x = (x^1, \quad, x^N) \in X \setminus \{0\}$ denote by $I(x)$ the subset of integers $i \in \{1, \quad, N\}$ for which $x^i = 0$

Let now ψ be an eigenfunction of H with eigenvalue $\mu < \inf \sigma_{ess}(H)$ Then

$$\int_X |\psi(x)|^2 \, e^{2(1-\varepsilon)\rho(x)} dx < \infty \tag{4 62}$$

for any $\varepsilon > 0$ where $\rho(x)$ is the geodesic distance from x to 0 in the Riemannian metric

$$ds^2 = (\Lambda_{I(x)} - \mu) \sum_{i=1}^{N} 2m_i |dx^i|^2$$

(Here we use the notation

$$|dx^i|^2 = (dx_1^i)^2 + \quad + (dx_\nu^i)^2 \qquad (4\ 63)$$

for $x^i = (x_1^i, \quad , x_\nu^i) \in \mathbf{R}^\nu$)

Proof As before we consider X to be an inner product space with an inner product given by (4 34) Let ω be a unit-vector in X It is readily seen that the subsystem Hamiltonian which in the notation of the theorem is denoted by $H_{I(\omega)}$ is the same as the Hamiltonian previously denoted by H_ω' Hence $K(\omega,p)$ is given by (4 58) which in the notation of the theorem takes the form

$$K(\omega,P) = \inf \sigma(H_{I(\omega)}) \text{ if } I(\omega) \text{ is not empty} \qquad (4\ 64)$$
$$K(\omega,P) = 0 \text{ if } I(\omega) \text{ is empty}$$

The conclusion of the theorem follows now by a straightforward application of Corollary 4 5 to P taking into account the special expression of the function $K(\omega,P)$ given in (4 64) ▪

Remark It follows from (2 22) and (3 16) that

$$\inf \sigma_{ess}(H) = \min K(\omega,P) \qquad (4\ 65)$$

where the minimum is taken over all unit-vectors ω We also have $\max K(\omega,P) = 0$ (see (4 28)) Hence combining (4 65) with (4 64) we obtain the following characterization of the bottom of the essential spectrum of H in terms of the bottoms of the spectra of the subsystem Hamiltonians H_I

$$\inf \sigma_{ess}(H) = \min_I \inf \sigma(H_I) \qquad (4\ 66)$$

where the minimum is taken over all proper subsets I of $\{1, \quad , N\}$ Formula (4 66) is the main part of the *HVZ* theorem for atomic type Schrodinger operators (see [35, Theorem XIII 17])

We now turn to the general N-body problem Consider a system of N particles with coordinates $x^i \in \mathbf{R}^\nu$, $i = 1, \quad , N$, masses $m_i \in \mathbf{R}_+$, and interacting potentials $v_{ij}(y)$ defined on $\mathbb{R}^\nu$ We assume that the functions v_{ij} $(1 \le i < j \le N)$ satisfy the following conditions

(i) $v_{ij} \in L^1_{loc}(\mathbb{R}^\nu)$ and $(v_{ij})_- \in M_{loc}(\mathbb{R}^\nu)$

(ii) $\lim_{|y|\to\infty} v_{ij}(y) = 0$

The Schrodinger differential operator $\widetilde{P}$ of the system is defined on the configuration space $\mathbb{R}^{\nu N}$ (generic point $x = (x^1, \quad , x^N)$, $x^i \in \mathbb{R}^\nu$) by

$$\widetilde{P} = -\sum_{i=1}^{N} (2m_i)^{-1} \Delta_i + \sum_{1\le i<j\le N} v_{ij}(x^i - x^j) = -\Delta + V(x) \tag{4 67}$$

where $\Delta = \sum_{i=1}^{N} (2m_i)^{-1}\Delta_i$ is the Laplace-Beltrami operator on $\mathbb{R}^{\nu N}$ with respect to the inner product

$$<x,y> = \sum_{i=1}^{N} 2m_i x^i y^i \tag{4 68}$$

and where $V(x) = \sum_{1\le i<j\le N} v_{ij}(x^i - x^j)$

The Schrodinger operator of the N-body system defined above is not quite the "right" operator to be considered in connection with eigenvalue problems since it has *no* eigenfunctions (this follows easily from the fact that $V(x+x_0) = V(x)$ for any x and any x_0 such that $x_0^1 = x_0^2 = \quad = x_0^N$) The operator we shall consider instead is the N-body Schrodinger operator with *center of mass removed* To define this operator introduce the subspace X of the configuration space defined by

$$X = \{x = (x^1, \quad , x^N) \in \mathbb{R}^{\nu N} \quad \sum_{i=1}^{N} m_i x^i = 0\} \tag{4 69}$$

Consider X as a $\nu(N-1)$ dimensional inner product space with inner product given by (4 68) Denote by Δ_X the Laplace-Beltrami operator in the inner product space X (Δ_X can also be described in the part of $\Delta = \sum_{i=1}^{N} (2m_i)^{-1}\Delta_i$ which acts on X) The N-body operator with center of mass removed is the operator P on X defined by

$$P = -\Delta_X + \sum_{1 \le i < j \le N} V_{ij}(x) = -\Delta_X + V(x)$$

where $V_{ij}(x) = v_{ij}(x^i - x^j)$ and $V = \Sigma\, V_{ij}$ are restricted to X. (Note that the functions V_{ij} on $\mathbf{R}^{\nu N}$ are constant along lines orthogonal to X).

In the following we shall also refer to P as the *reduced* N-body Schrödinger operator.

Let Π_{ij} , $1 \le i < j \le N$, be the projections in $\mathbf{R}^{\nu N}$ defined by (4.35). As was noted before Π_{ij} are orthogonal projections in the inner product (4.68). It is also readily checked that Ran $\Pi_{ij} \subset X$ and that

$$V_{ij}(\Pi_{ij} x) = V_{ij}(x) \quad \text{for} \quad x \in X\,,\; 1 \le i < j \le N.$$

It is thus seen that P verifies the conditions of Lemma 4.8 and that we are in a position to apply to it Theorem 4.9.

When applying Theorem 4.9 we must decide when $\Pi_{ij}\omega = 0$, $\omega = (\omega', \ldots, \omega^N) \in X$. Clearly this happens if and only if $\omega^i = \omega^j$. Therefore the operators P_ω which figure in the theorem are defined in the present case by

$$P_\omega = -\Delta_X + \sum_{\substack{i<j \\ \omega^i = \omega^j}} V_{ij},$$

$\omega = (\omega^1, \ldots, \omega^N)$ a given unit-vector in X. Observe that P_ω can be described as the Schrödinger operator of N-particle system obtained by a partition of the N particles $\{1, \ldots, N\}$ into clusters defined as equivalence classes by the equivalence relation: $i \sim j$ if and only if $\omega^i = \omega^j$.

Applying Theorem 4.9 we obtain the following exponential decay result.

4.13 Theorem: *Let*

$$P = -\Delta_X + \sum_{1 \le i < j \le N} v_{ij}(x^i - x^j) = -\Delta_X + V(x) \tag{4.70}$$

be the reduced N-body Schrödinger operator associated with a system of N particles with coordinates $x^i \in \mathbf{R}^\nu$, masses $m_i \in \mathbf{R}_+$ and interacting potentials

$v_{ij} \in L^1_{loc}(\mathbf{R}^\nu)$. *P acts on functions on the subspace* $X \subset \mathbf{R}^{\nu N}$ defined by $X = \{x = (x^1, \ldots, x^N) : \sum_{i=1}^{N} m_i x^i = 0\}$. *Here* Δ_X *denotes the Laplace-Beltrami operator on X with respect to the restriction to X of the inner product defined by* (4.68). *We assume also that the functions* $v_{ij}(y)$ *on* $\mathbf{R}^\nu$ *verify the following conditions:* $(v_{ij})_- \in M_{loc}(\mathbf{R}^\nu)$ *and* $\lim_{|y|\to\infty} v_{ij}(y) = 0$. *We consider the functions* $v_{ij}(x^i - x^j)$ *and* $V(x) = \sum_{i<j} v_{ij}(x^i - x^j)$ *as functions defined on X.*

Let H be the self-adjoint realization of P in $L^2(X)$ *and let* ψ *be an eigenfunction of H with eigenvalue* $\mu < \inf \sigma_{ess}(H)$. *Then*

$$\int_X |\psi(x)|^2 e^{2(1-\varepsilon)\rho(x)} dx < \infty \tag{4.71}$$

for any $\varepsilon > 0$ *where* $\rho(x)$ *denotes the geodesic distance from x to the origin in X in the Riemannian metric:*

$$ds^2 = c(x) \sum_{i=1}^{N} 2m_i |dx^i|^2, \tag{4.72}$$

($|dx^i|^2$ *is given by* (4.63)) *where* $c(x)$ *is a positive lower semicontinuous function on* $X \setminus \{0\}$ *defined as follows. With every unit-vector* ω *in X associate a cluster-partition Schrödinger operator:*

$$P_\omega = -\Delta_X + \sum_{\substack{i<j \\ \omega^i = \omega^j}} v_{ij}(x^i - x^j). \tag{4.73}$$

Denote by H_ω *the self-adjoint realization of* P_ω *in* $L^2(X)$. Set

$$\Sigma_\omega = \inf \sigma(H_\omega).$$

Then

$$c(x) = \Sigma_{x/|x|}.$$

The exponential decay results for eigenfunctions proved in Theorem 4.13, Theorem 4.12 and Theorem 4.9 could easily be extended to more general solutions of the differential equation $(P-\mu)\psi = 0$ where ψ is defined only in some neighborhood Ω of infinity, ψ need not belong to $L^2(\Omega)$ but satisfy instead a weaker growth restriction, and μ may be complex. For example, we give the following generalization of Theorem 4.13.

4.14 Theorem: *Let $P = -\Delta_X + V$ be the reduced N- body Schrödinger operator defined by* (4.70), *satisfying the same conditions as in Theorem 4.13. For every unit-vector ω in X set*

$$\Lambda_\omega = \Lambda(P_\omega) = \inf\{\frac{(P_\omega \varphi, \varphi)}{\|\varphi\|} : \varphi \in C_0^\infty(X)\, , \varphi \neq 0\}$$

where P_ω is defined by (4.73).

Let Ω be an unbounded open set in X such that $X \setminus \Omega$ is compact. Let $\psi \in H^1_{loc}(\Omega)$ satisfy $V\psi \in L^1_{loc}(\Omega)$ and $P\psi = \mu\psi$ in Ω for some $\mu \in \mathbb{C}$ with $\mathrm{Re}\, \mu < \Sigma(P)$. *Let $\rho(x)$ denote the geodesic distance from x to the origin in X in the Riemannian metric*

$$ds^2 = (\Lambda_{x/|x|} - \mathrm{Re}\mu) \sum_{i=1}^{N} 2m_i \, |dx^i|^2 .$$

Suppose that

$$\int_\Omega |\psi(x)|^2 e^{-2(1-\delta)\rho(x)} dx < \infty$$

for some $\delta > 0$. Then

$$\int_\Omega |\psi(x)|^2 e^{2(1-\varepsilon)\rho(x)} dx < \infty$$

for any $\varepsilon > 0$.

Proof: By Lemma 4.8 we have

$$K(\omega;P) = \Lambda_\omega \tag{4.74}$$

for all unit-vectors ω. The conclusion of the theorem follows now by a straightforward application of Theorem 4.4, using (4.74) and noting that $K(\omega;P) \leq 0$ in view of (4.28). ▪

Chapter 5 Pointwise Exponential Bounds

In this chapter we will derive pointwise bounds for solutions of the equation $Au+qu=0$ in a bounded subset of $\mathbf{R}^n$. When combined with the results of Chapter 4 these bounds will give pointwise exponential decay bounds for eigenfunctions of the N-body Schrödinger Hamiltonians considerd in Theorem 4.13, as well as pointwise exponential decay bounds for eigenfunctions of the atomic type Schrödinger Hamiltonians considered in Theorem 4.12 and for eigenfunctions of the general multiparticle type Schrödinger operators considered in Theorem 4.9.

Pointwise bounds for solutions of the equation $Au=0$ were derived in the well known papers by De Giorgi [9] and Nash [29]. A simple and elegant derivation of these results is due to Moser [28]. An extension of the De Giorgi-Nash results to solutions of more general equations including equations of the form $Au+qu=0$ with q real were given by Stampacchia [39, 40]. Other extensions were given by Ladyzhenskaya and Ural'tseva [23] and by Trudinger [42].

The main theorem in this chapter gives pointwise bounds for solutions of equations $Au+qu=0$ with q verifying conditions which are very close to the general conditions imposed on q in Chapter 1. The result we establish is of the type proved by Stampacchia in the papers mentioned above but we shall allow a more general class of "potentials" q. In particular we shall allow q and u to be complex.

When deriving the interior pointwise estimates we shall assume without loss of generality that the equation $Au+qu=0$ is given in the unit ball $B=B(0;1)$ in $\mathbf{R}^n$. We will work with complex valued q such that $q_- \in M_\delta(B)$ for some $\delta>0$ where as before we set

$$q_- = \max(0, -\mathrm{Re}\, q).$$

This assumption is equivalent to the assumption $q_- \in L^1(B)$ and

$$\sup_{x \in B} \int_B q_-(y)|y-x|^{2-n-\delta} dy < \infty$$

for some $\delta > 0$. Note in particular that if $q_- \in L^p(B)$ for some $p > n/2$ $(n \geq 2)$ then $q_- \in M_\delta(B)$ for some $\delta > 0$.

In view of our assumption on q it follows by Lemma 0.2 that there is a constant c_1 and a number $\vartheta \in (0,1)$ such that for all $\varphi \in C_0^\infty(B)$:

$$\|q_-^{1/2}\varphi\|_{L^2(B)} \leq c_1 \|\Lambda^\vartheta \varphi\|_{L^2(B)} \tag{5.1}$$

where $\Lambda = (1-\Delta)^{1/2}$.

5.1 Theorem: *Suppose that*

(i) $a^{ij}(x)$, $1 \leq i,j \leq n$, *are real measurable functions on* B *satisfying* $a^{ij}(x) = a^{ji}(x)$ *and*

$$c_2^{-1}|\xi|^2 \leq \sum_{i,j=1}^{n} a^{ij}(x)\xi_i\xi_j \leq c_2|\xi|^2 \tag{5.2}$$

for all $\xi \in \mathbb{R}^n$ *and* $x \in B$, c_2 *a positive constant.*

(ii) q *is a complex valued function in* $L^1_{loc}(B)$ *with* q_- *satisfying* (5.1) *for all* $\varphi \in C_0^\infty(B)$ *with some* $\vartheta \in (0,1)$.

Suppose u *is a weak solution of the equation*

$$Au + qu \equiv -\sum_{i,j=1}^{n} \partial_j a^{ij}\partial_i u + qu = 0 \tag{5.3}$$

in the sense that $u \in L^2(B) \cap H^1_{loc}(B)$, $qu \in L^1_{loc}(B)$, *and for each* $\varphi \in C_0^\infty(B)$

$$\int_B (\Delta_A u \cdot \Delta_A \varphi + qu\varphi)dx = 0 \tag{5.4}$$

(we use the notation (1.3)*). Then* $u \in L^\infty_{loc}(B)$ *and for each ball* $B_a = B(0;a)$ *with* $0 < a < 1$, *there is a constant* $D = D(c_1,c_2,a)$ *(where* c_1 *and* c_2 *are given in* (5.1) *and* (5.2)*) such that*

$$\operatorname*{ess\,sup}_{x \in B_a} |u(x)| \leq D\|u\|_{L^2(B)}. \tag{5.5}$$

Before we proceed with the somewhat long proof of Theorem 5.1 we shall apply it to derive exponential decay pointwise bounds for eigenfunctions of the atomic type Schrödinger operators considered in Theorem 4.12 and for eigenfunctions of the N-body Schrödinger operators considered in Theorem 4.13.

5.2 Theorem: *Let*

$$P = \sum_{i=1}^{N}(-(2m_i)^{-1}\Delta_i + v_{oi}(x^i)) + \sum_{1\le i<j\le N} v_{ij}(x^i - x^j)$$

be the atomic type Schrödinger operator introduced in Theorem 4.12, P acts on functions defined on $X = \mathbb{R}^{\nu N}$. Suppose in addition to the conditions imposed on the v_{ij} *in Theorem 4.12 that $(v_{oi})_- \in M_\delta(\mathbb{R}^\nu)$ for some $\delta > 0$, $i = 1, \ldots, N$.*

Let H be the self-adjoint realization of P in $L^2(X)$. Let $\psi(x)$ be an eigenfunction of H with eigenvalue $\mu < \inf \sigma_{ess}(H)$. Then for every $\varepsilon > 0$ there is a constant C_ε such that

$$|\psi(x)| \le C_\varepsilon e^{-(1-\varepsilon)\rho(x)} \quad a.e. \quad on \quad X$$

where $\rho(x)$ is the geodesic distance function from x to the origin defined in Theorem 4.12.

5.3 Theorem: *Let*

$$P = -\Delta_X + \sum_{1\le i<j\le N} v_{ij}(x^i - x^j) = -\Delta_X + V(x)$$

be the reduced N-body Schrödinger operator introduced in Theorem 4.13, P acts on functions defined on the subspace $X \subset \mathbb{R}^{\nu(N-1)}$ defined by (4.69). Suppose also that $(v_{ij})_- \in M_\delta(\mathbb{R}^\nu)$ for some $\delta > 0$, $1 \le i < j \le N$.

Let H be the self-adjoint realization of P in $L^2(X)$. Let $\psi(x)$ be an eigenfunction of H with eigenvalue $\mu < \inf \sigma_{ess}(H)$. Then for every $\varepsilon > 0$ there is a constant C_ε such that

$$|\psi(x)| \le C_\varepsilon e^{-(1-\varepsilon)\rho(x)} \quad a.e. \quad on \quad X \tag{5.6}$$

where $\rho(x) = \rho(x,0)$, $\rho(x,y)$ denoting the geodesic distance from x to y in the

Riemannian metric (4 72) *as defined in Theorem 4 13*

Remark The following less general but simpler conditions on the potentials $v_{ij}(y)$ (on $\mathbb{R}^\nu$) are sufficient to ensure the validity of Theorem 5 2

(a) $v_{ij} \in L^1_{loc}(\mathbb{R}^\nu)$ and $\lim_{|y|\to\infty} v_{ij}(y) = 0$ for all i, j,

and

(b) $(v_{oi})_- \in L^p_{loc}(\mathbb{R}^\nu)$ with $p > \nu/2$ for $i = 1, \quad , N$

Similarly the conditions of Theorem 5 3 hold if the v_{ij} verify (a) as above and

(b)' $(v_{ij})_- \in L^p_{loc}(\mathbb{R}^\nu)$ with $p > \nu/2$ for $1 \le i < j \le N$

We shall only give the proof of Theorem 5 3 since the proof of Theorem 5 2 follows exactly the same lines (Both theorems follow by combining the proper L^2 estimate established in Chapter 4 with the pointwise estimate of Theorem 5 1)

Proof of Theorem 5 3 Since $(v_{ij})_- \in M_\delta(\mathbb{R}^\nu)$ it follows as in the proof of Lemma 4 7 that $V_{ij}(x) = v_{ij}(x^i - x^j)$ is in $M_\delta(X)$ where we identify X with $\mathbb{R}^{\nu(N-1)}$ Since $V_- \le \sum_{1\le i<j\le N} (V_{ij})_-$ it follows that $V_- \in M_\delta(X)$ and hence, in view of Lemma 0 2, the estimate

$$\| V_-^{1/2} \varphi \|_{L^2(X)} \le c_1 \| \Lambda^\vartheta \varphi \|_{L^2(X)}$$

holds for all $\varphi \in C_0^\infty(X)$ for some $\vartheta \in (0,1)$

Applying Theorem 5 1 to the eigenfunction ψ it follows that for every fixed $x_0 \in X$

$$\operatorname*{ess\,sup}_{B(x_0\, 1/2)} |\psi(x)| \le D \|\psi\|_{L^2(B(x_0\, 1))},$$

where D is independent of x_0 We thus have for $\varepsilon > 0$ and $\rho = \rho(x)$

$$\operatorname*{ess\,sup}_{B(x_0\, 1/2} |\psi e^{(1-\varepsilon)\rho}| \le D \|\psi\|_{L^2(B(x_0\, 1)} \{ \sup_{x \in B(x_0\, 1)} e^{(1-\varepsilon)\rho(x)} \}$$

$$\leq D\,||\psi e^{(1-\varepsilon)\rho}||_{L^2(B(x_0;1))}\{\sup_{x,y\in B(x_0;1)} e^{(1-\varepsilon)(\rho(x)-\rho(y))}\}$$

$$\leq D\,||\psi e^{(1-\varepsilon)\rho}||_{L^2(X)}\{\sup_{x,y\in B(x_0;1)} e^{(1-\varepsilon)\rho(x,y)}\}$$

$$\leq D'\,||\psi e^{(1-\varepsilon)\rho}||_{L^2(X)},$$

where the last inequality follows because $\rho(x,y)\leq c\,|x-y|$. Since x_0 is arbitrary the result follows from Theorem 4.13. ▪

The remainder of this chapter is devoted to the proof of Theorem 5.1. We start by proving a lemma which will allow us to reduce the proof of Theorem 5.1 to the case where u is a real subsolution of the equation $A'v - q_-v = 0$. The lemma is an extension of Kato's distributional inequality [20; Lemma A] to the case of complex solutions of elliptic equations in divergence form with measurable bounded coefficients. (In this connection we mention that Simon [38] has given an abstract version of Kato's inequality which holds for a class of positive operators in a Hilbert space.)

5.4 Lemma: *Let* $A = -\sum_{i,j}\partial_j a^{ij}(x)\partial_i$ *be the elliptic operator introduced in Theorem* 5.1 *and let* Ω *be an open set contained in* B. *Let* $u(x)$ *be a complex valued function in* $H^1_{loc}(\Omega)$. *Suppose that* $Au = f$, *for some* $f \in L^1_{loc}(\Omega)$, *in the sense that*

$$\int_\Omega \nabla_A u\cdot\nabla_A\psi\, dx = \int_\Omega f\,\psi\, dx \tag{5.7}$$

for all $\psi \in C_0^\infty(\Omega)$. *Then* $|u| \in H^1_{loc}(\Omega)$ *and*

$$A|u| \leq \mathrm{Re}[f\,(\mathrm{sgn}\,\bar{u})]$$

in the sense that

$$\int_\Omega \nabla_A|u|\cdot\nabla_A\varphi\, dx \leq \int_\Omega \mathrm{Re}[f\,(\mathrm{sgn}\,\bar{u})]\varphi\, dx \tag{5.8}$$

for all non-negative functions φ *in* $C_0^\infty(\Omega)$.

Remark: The function $\mathrm{sgn}\, z : \mathbf{C}\to\mathbf{C}$ is defined by $\mathrm{sgn}\, z = z/|z|$ for $z\neq 0$, $\mathrm{sgn}\, 0 = 0$.

Proof: For any $\varepsilon > 0$ define $u_\varepsilon = (|u|^2 + \varepsilon^2)^{1/2}$. We claim that u_ε and $\bar{u}/u_\varepsilon$ are in $H^1_{loc}(\Omega)$ with

$$\partial_i u_\varepsilon = \frac{1}{2}(\bar{u}/u_\varepsilon)\partial_i u + \frac{1}{2}(u/u_\varepsilon)\partial_i \bar{u} = \mathrm{Re}[(\bar{u}/u_\varepsilon)\partial_i u], \tag{5.9}$$

$$\partial_i(\bar{u}/u_\varepsilon) = (\partial_i \bar{u})/u_\varepsilon - \bar{u}(\partial_i u_\varepsilon)/u_\varepsilon^2. \tag{5.10}$$

To see this we will use the product rule $\partial_i(fg) = g\,\partial_i f + f\,\partial_i g$ for $f, g \in H^1_{loc}(\Omega)$) and the chain rule (Lemma 5.5). The product rule is easily proved by an approximation argument. To see (5.9) let F_1 be in $C^1(\mathbb{R})$ with bounded derivative and $F_1(x) = (x+\varepsilon^2)^{1/2}$ for $x \geq 0$. Then $u_\varepsilon = F_1 \circ |u|^2$ and the chain rule can be applied.

Similarly to compute the (distributional) derivatives of u_ε^{-1} we introduce a function F_2 in $C^1(\mathbb{R})$ with bounded derivative such that $F_2(x) = x^{-1}$ if $x \geq \varepsilon$. Then $u_\varepsilon^{-1} = F_2 \circ u_\varepsilon$ so that applying the chain rule and product rule results in (5.10).

By combining (5.9) and (5.10) one obtains the following formula which we will make use of later on

$$\partial_i(\bar{u}/u_\varepsilon) = (\partial_i \bar{u})/u_\varepsilon - \frac{1}{2}(\bar{u}^2 \partial_i u + \bar{u}u\,\partial_i \bar{u})/u_\varepsilon^3. \tag{5.11}$$

Since $|u/u_\varepsilon| < 1$ and since $u(x)/u_\varepsilon(x) \to \mathrm{sgn}\, u(x)$ almost everywhere in Ω, it follows from (5.9) by the dominated convergence theorem that as $\varepsilon \to 0$

$$\partial_i u_\varepsilon \to \frac{1}{2}(\mathrm{sgn}\,\bar{u})\partial_i u + \frac{1}{2}(\mathrm{sgn}\, u)\partial_i \bar{u} \quad in \quad L^1_{loc}(\Omega). \tag{5.12}$$

Since $u_\varepsilon \to |u|$ in $L^1_{loc}(\Omega)$ as $\varepsilon \to 0$ it follows from (5.12) that u_ε converges to a limit in $H^1_{loc}(\Omega)$. This implies that $|u| \in H^1_{loc}(\Omega)$ and that

$$u_\varepsilon \to |u| \quad in \quad H^1_{loc}(\Omega) \quad as \quad \varepsilon \to 0. \tag{5.13}$$

Now, let φ be a non-negative function in $C_0^\infty(\Omega)$ and consider the function

$$\nabla_A u_\varepsilon \cdot \nabla_A \varphi = \sum_{i,j} a^{ij}(x)\partial_i u_\varepsilon(x)\partial_j \varphi(x). \tag{5.14}$$

Inserting the expression for $\partial_i u_\varepsilon$ given in (5.9) into (5.14) it follows that almost everywhere in Ω

$$\nabla_A u_\varepsilon \cdot \nabla_A \varphi = \mathrm{Re}(\frac{\bar{u}}{u_\varepsilon}\sum_{i,j} a^{ij}\partial_i u \partial_j \varphi)$$

$$= \mathrm{Re}(\frac{\bar{u}}{u_\varepsilon}\nabla_A u \cdot \nabla_A \varphi) = \mathrm{Re}(\nabla_A u \cdot \nabla_A (\frac{\bar{u}}{u_\varepsilon}\varphi)$$

$$- \varphi\, \mathrm{Re}(\nabla_A u \cdot \nabla_A (\frac{\bar{u}}{u_\varepsilon})). \tag{5.15}$$

We shall show that

$$\mathrm{Re}(\nabla_A u \cdot \nabla_A (\frac{\bar{u}}{u_\varepsilon})) \geq 0 \quad a.e. \quad in \quad \Omega, \tag{5.16}$$

which when combined with (5.15) will prove that

$$\nabla_A u_\varepsilon \cdot \nabla_A \varphi \leq \mathrm{Re}(\nabla_A u \cdot \nabla_A (\frac{\bar{u}}{u_\varepsilon}\varphi)) \tag{5.17}$$

a.e. in Ω for any non-negative function φ in $C_0^\infty(\Omega)$.

To prove (5.16) observe that in view of (5.11)

$$\mathrm{Re}(\nabla_A u \cdot \nabla_A (\frac{\bar{u}}{u_\varepsilon}) = \mathrm{Re}\ \sum_{i,j} a^{ij}\partial_i u \partial_j (\frac{\bar{u}}{u_\varepsilon})$$

$$= u_\varepsilon^{-1}\sum_{i,j}\partial_i u \partial_j \bar{u} - \frac{1}{2}\mathrm{Re}(\bar{u}^2 u_\varepsilon^{-3}\sum_{i,j} a^{ij}\partial_i u \partial_j u)$$

$$- \frac{1}{2}\mathrm{Re}(\bar{u}u u_\varepsilon^{-3}\sum_{i,j}\partial_i u \partial_j \bar{u}) = u_\varepsilon^{-1}|\nabla_A u|^2$$

$$- \frac{1}{2}\mathrm{Re}(\bar{u}^2 u_\varepsilon^{-3}\nabla_A u \cdot \nabla_A u) - \frac{1}{2}\mathrm{Re}(\bar{u}u u_\varepsilon^{-3}|\nabla_A u|^2). \tag{5.18}$$

It follows from (5.18) (since $|\nabla_A u \cdot \nabla_A u| \leq |\nabla_A u|^2$ and $|u/u_\varepsilon| < 1$) that (a.e. in Ω)

$$\mathrm{Re}(\nabla_A u \cdot \nabla_A (\frac{\bar{u}}{u_\varepsilon}))$$

$$\geq u_\varepsilon^{-1}(|\nabla_A u|^2 - \frac{1}{2}|\bar{u}^2 u_\varepsilon^{-2}\nabla_A u \cdot \nabla_A u| - \frac{1}{2}|u|^2 u_\varepsilon^{-2}|\nabla_A u|^2)$$

$$\geq u_\varepsilon^{-1}(|\nabla_A u|^2 - \frac{1}{2}|\nabla_A u|^2 - \frac{1}{2}|\nabla_A u|^2) \geq 0,$$

which proves (5.16).

Integrating both sides of the inequality (5.17) over Ω we find that

$$\int_\Omega \nabla_A u_\varepsilon \cdot \nabla_A \varphi \, dx \le \mathrm{Re} \int_\Omega \nabla_A u \cdot \nabla_A \left(\frac{\bar{u}}{u_\varepsilon} \varphi\right) dx. \tag{5.19}$$

Using our assumption that $Au = f \in L^1_{loc}(\Omega)$ in the weak sense (of (5.7)) it follows readily that

$$\int_\Omega \nabla_A u \cdot \nabla_A \left(\frac{\bar{u}}{u_\varepsilon} \varphi\right) dx = \int_\Omega f \frac{\bar{u}}{u_\varepsilon} \varphi dx. \tag{5.20}$$

Indeed, to prove (5.20) just note that since $\bar{u}/u_\varepsilon \in H^1_{loc}(\Omega)$ and $|\bar{u}/u_\varepsilon| < 1$, there exists a sequence of functions $\psi_m \in C_0^\infty(\Omega)$, $m = 1,2,\cdots$, with $|\psi_m| \le 1$, such that $\psi_m \to \bar{u}/u_\varepsilon$ in $H^1_{loc}(\Omega)$. (One can take $\psi_m = \bar{\varphi}_m(|\varphi_m|^2 + \varepsilon)^{-1/2}$ where $\{\varphi_m\}$ is any sequence of functions in $C_0^\infty(\Omega)$ such that $\varphi_m \to u$ in $H^1_{loc}(\Omega)$ and such that $\varphi_m(x) \to u(x)$ for almost all x.) Applying (5.7) with $\psi = \psi_m \varphi$ and then letting $m \to \infty$, the relation (5.20) follows by Lebesgue's dominated convergence theorem.

Combining (5.19) with (5.20) we get

$$\int_\Omega \nabla_A u_\varepsilon \cdot \nabla_A \varphi dx \le \mathrm{Re} \int_\Omega f \frac{\bar{u}}{u_\varepsilon} \varphi dx. \tag{5.21}$$

Since $u_\varepsilon \to |u|$ in $H^1_{loc}(\Omega)$ as $\varepsilon \to 0$, by (5.13), and since $\bar{u}(x)/u_\varepsilon(x) \to \mathrm{sgn}\, \bar{u}(x)$ as $\varepsilon \to 0$ for $x \in \Omega$, it follows from (5.21) upon letting $\varepsilon \to 0$ (by the dominated convergence theorem) that

$$\int_\Omega \nabla_A |u| \cdot \nabla_A \varphi dx \le \int_\Omega \mathrm{Re}[f(\mathrm{sgn}\, \bar{u})] \varphi \, dx.$$

This yields (5.8) and proves the lemma. ▪

Before going on to prove Theorem 5.1 we will need a few more technical results:

5.5 Lemma (chain rule): *Suppose $G : \mathbb{R} \to \mathbb{R}$ is continuous and there exists a finite set S such that $G \in C^1(\mathbb{R} \setminus S)$ and for all $t \in \mathbb{R} \setminus S$, $|G'(t)| \le M$. Suppose u is a real valued function in $L^1_{loc}(B)$ with distributional derivatives $\partial_i u$ in $L^1_{loc}(B)$. Then $G \circ u$ and $\partial_i G \circ u$ are in $L^1_{loc}(B)$ and the following formula* holds almost everywhere in B

$$\partial_i(G\circ u) = \begin{cases} G'(u)\partial_i u & \text{if } u \notin S \\ 0 & \text{if } u \in S \end{cases}$$

In addition suppose $c \in \mathbb{R}$ *and* $S_c = \{x \quad u(x) = c\}$ *Then*

$$\chi_{S_c}\partial_i u = 0$$

almost everywhere Here χ_D *denotes the characteristic function of the set* D $\chi_D(x) = 1$ *if* $x \in D$, $\chi = 0$ *otherwise*

A short proof of this lemma can be found in [12, §7 4]

5.6 Corollary: *Suppose* u *is real valued and in* $L^p_{loc}(B)$ *for some* $p \in [1,\infty)$ *and* $\partial_i u \in L^1_{loc}(B)$ *If in addition either*

(i) $|u|^{p-1}\partial_i u \in L^1_{loc}(B)$

or

(ii) $\partial_i |u|^p \in L^1_{loc}(B)$

then

$$\partial_i |u|^p = p\,|u|^{p-1}(\operatorname{sgn} u)\partial_i u \tag{5 22}$$

Proof: Let $G_{N,p}(t) = |t|^p$ if $|t| < N$ and $G_{N,p}(t) = N^p$ if $|t| \geq N$ Let $S_N = \{x \quad |u(x)| < N\}$ If u and $\partial_i u$ are in $L^1_{loc}(B)$, then by Lemma 5 5

$$\partial_i(G_{N,p}\circ u) = \chi_{S_N} p\,(\operatorname{sgn} u)|u|^{p-1}\partial_i u \tag{5 23}$$

Suppose $u \in L^p_{loc}(B)$, $\partial_i u \in L^1_{loc}(B)$, $|u|^{p-1}\partial_i u \in L^1_{loc}(B)$ and $\varphi \in C_0^\infty(B)$ Then

$$\begin{aligned} -\int_B (\partial_i\varphi)|u|^p\,dx &= \lim_{N\to\infty} -\int_B (\partial_i\varphi)G_{N,p}\circ u\,dx \\ &= \lim_{N\to\infty}\int_B \varphi\chi_{S_N} p\,(\operatorname{sgn} u)|u|^{p-1}\partial_i u\,dx \\ &= \int_B \varphi p\,(\operatorname{sgn} u)|u|^{p-1}\partial_i u\,dx \end{aligned}$$

where the first and third equalities follow from the dominated convergence theorem. This shows that (i) implies (5.22). Now suppose $u \in L^p_{loc}(B)$ and $\partial_i |u|^p \in L^1_{loc}(B)$. Then

$$G_{N^p,1} \circ |u|^p = G_{N,p} \circ u$$

so that by (5.23)

$$\chi_{S_N}(\operatorname{sgn} |u|^p)\partial_i |u|^p = \chi_{S_N} p(\operatorname{sgn} u)|u|^{p-1}\partial_i u.$$

Since $\partial_i |u|^p = 0$ almost everywhere on the set where $u = 0$, letting N tend to infinity gives (5.22). Thus (ii) implies (5.22). ∎

We will also need two consequences of the bound (5.1).

5.7 Lemma: *Suppose $\vartheta \in (0,1)$. Then there is a constant* c *and a number $r \in (1,2)$ such that for all $\varphi \in C_0^\infty(\mathbb{R}^n)$ and all $\varepsilon > 0$*

$$\|\Lambda^\vartheta \varphi\|_{L^2(\mathbb{R}^n)} \le c(\|\nabla\varphi\|_{L^r(\mathbb{R}^n)} + \|\varphi\|_{L^r(\mathbb{R}^n)}), \tag{5.24}$$

$$\|\Lambda^\vartheta \varphi\|_{L^2(\mathbb{R}^n)} \le \varepsilon\|\nabla\varphi\|_{L^2(\mathbb{R}^n)} + c(\varepsilon^{-\vartheta/(1-\vartheta)} + \varepsilon)\|\varphi\|_{L^2(\mathbb{R}^n)}. \tag{5.25}$$

Lemma 5.7 is proved in Appendix 4. We now turn to the

Proof of Theorem 5.1: Suppose that u is a complex valued function in $H^1_{loc}(B)$ which satisfies the differential equation $Au + qu = 0$ in the weak sense (5.4). Applying Lemma 5.4 (with $f = -qu \in L^1_{loc}(B)$) it follows that $|u| \in H^1_{loc}(B)$ and that

$$A|u| \le \operatorname{Re}(-qu(\operatorname{sgn} \bar{u})) = \operatorname{Re}(-q)|u| \le q_-|u| \tag{5.26}$$

in the sense that

$$\int_B \nabla_A |u| \cdot \nabla_A \varphi \, dx - \int_B q_-|u|\varphi \, dx \le 0 \tag{5.27}$$

for all non-negative functions φ in $C_0^\infty(B)$. By a limiting argument (5.27) holds for all non-negative functions φ in $L_0^\infty(B) \cap H^1_{loc}(B)$. We now choose a particular φ. Define G_N and u_N as follows:

$$G_N(t) = \begin{cases} |t| & \text{if } |t| < N \\ N & \text{if } |t| \ge N, \end{cases}$$

$$u_N = G_N \circ |u|.$$

We also write $S_N = \{x : |u(x)| < N\}$. Then Lemma 5.5 implies

$$\partial_i u_N = \chi_{S_N} \partial_i |u|. \tag{5.28}$$

Let $\zeta \in C_0^\infty(B)$ be non-negative and fix $p \geq 2$. Define

$$\varphi = \zeta^2 u_N^{p-1}.$$

The function φ can also be written in the form

$$\varphi = \zeta^2 G_{N,p} \circ |u|$$

where

$$G_{N,p}(t) = \begin{cases} |t|^{p-1} & \text{if } |t| < N, \\ N^{p-1} & \text{if } |t| \geq N. \end{cases}$$

With φ written in this form Lemma 5.5 can be applied to give

$$\partial_i \varphi = (2\zeta \partial_i \zeta) u_N^{p-1} + \zeta^2 (p-1) u_N^{p-2} \partial_i u_N \tag{5.29}$$

where we have also used (5.28) in deriving (5.29). Substituting this φ in (5.27) we get

$$(p-1)\int_B |\nabla_A u_N|^2 \zeta^2 u_N^{p-2} dx + 2\int_B \zeta u_N^{p-1} \nabla_A |u| \cdot \nabla_A \zeta \, dx$$
$$- \int_B q_- |u| \zeta^2 u_N^{p-1} dx \leq 0. \tag{5.30}$$

Using (5.28) we have

$$u_N^{p-2} |\nabla_A u_N|^2 = \chi_{S_N} |u|^{p-2} |\nabla_A |u||^2,$$

so that substituting this in (5.30) we find

$$(p-1)\int_{S_N} \zeta^2 |\nabla_A |u||^2 |u|^{p-2} dx \leq 2\int_B \zeta |u_N|^{p-1} |\nabla_A |u| \cdot \nabla_A \zeta| dx$$
$$+ \int_B q_- |u| \zeta^2 u_N^{p-1} dx.$$

Letting $N \to \infty$ in the last inequality, using the monotone convergence theorem, we thus get

$$(p-1)\int_B \zeta^2 |\nabla_A |u||^2 |u|^{p-2} dx \leq 2\int_B \zeta |u|^{p-1} |\nabla_A |u| \cdot \nabla_A \zeta| dx$$

$$+ \int_B q_- \zeta^2 |u|^p \, dx. \tag{5.31}$$

We now fix some $p_0 \geq 2$ and assume that $|u|^{p_0/2} \in H^1_{loc}(B)$. Note that we know already that this is true for $p_0 = 2$. We will show that $|u|^{p_0/2} \in H^1_{loc}(B)$ implies $|u|^{p/2} \in H^1_{loc}(B)$ for $p = \eta p_0$ where $\eta > 1$ is a number which will be determined. As we shall see, this will allow us to conclude that $|u|^{p/2} \in H^1_{loc}(B)$ for all $p \geq 2$. The number η will be chosen so that the right side of (5.31) is finite.

Consider the first term on the right side of (5.31). We have by Schwarz's inequality

$$\int_B \zeta |u|^{p-1} |\nabla_A |u| \cdot \nabla_A \zeta| \, dx \leq \left(\int_B \zeta^2 |u|^{p_0-2} |\nabla_A |u||^2 dx\right)^{1/2} \left(\int_B |\nabla_A \zeta|^2 |u|^{2p-p_0} dx\right)^{1/2}$$

$$= 2p_0^{-1} \left(\int_B \zeta^2 |\nabla_A |u|^{p_0/2}|^2 dx\right)^{1/2} \left(\int_B |\nabla_A \zeta|^2 |u|^{2p-p_0} dx\right)^{1/2} \tag{5.32}$$

where we have used Corollary 5.6 and the induction assumption, $|u|^{p_0/2} \in H^1_{loc}(B)$, to write $\partial_i |u|^{p_0/2} = (p_0/2)|u|^{(p_0/2)-1} \partial_i |u|$. To choose p so that the right side of (5.32) is finite we make use of Sobolev's inequality [12; §7.7] which states that if $n \geq 3$ then for all $v \in H^1(B)$

$$\|v\|_{L^{2^*}(B)} \leq c(\|\nabla v\|_{L^2(B)} + \|v\|_{L^2(B)}) \tag{5.33}$$

where, if $n \geq 3$, $2^* = 2n/(n-2)$ and c is some constant depending only on n. If $n = 2$ then (5.33) holds if 2^* stands for any number $r > 2$ (c depending on r), while for $n = 1$ (5.33) holds with $2^* = \infty$. In order not to consider special cases we define in the following $2^* = 3$ for $n = 1$ or 2. With this definition (5.33) holds for all n.

It follows from (5.33) and the hypothesis $|u|^{p_0/2} \in H^1_{loc}(B)$ that

$$|u|^{\frac{2^*}{2}(p_0/2)} \in L^1_{loc}(B) \tag{5.34}$$

and thus the right side of (5.32) is finite if $p - (p_0/2) \leq (2^*/2)(p_0/2)$, i.e., if $p \leq p_0(2^*+2)/4$.

We now consider the expression $\int_B q_-\zeta^2|u|^p dx$. At this stage we impose on ζ the restriction that it is of the form $\zeta = \zeta_1^2$ where ζ_1 is a non-negative function in $C_0^\infty(B)$. From Lemma 5.7 and (5.1) we have for some $r \in (1,2)$

$$||q_-^{1/2}\varphi||_{L^2(B)} \le c'(||\nabla\varphi||_{L^r(B)} + ||\varphi||_{L^r(B)}) \qquad (5.35)$$

for all $\varphi \in C_0^\infty(B)$ where c' is a constant. This easily extends to $\varphi \in L_0^\infty(B)$ with $\nabla\varphi \in L^\infty(B)$ and thus we can substitute $\varphi = |\psi|^{2/r}$ into (5.35) with $\psi \in C_0^\infty(B)$, and real valued. Using Corollary 5.6 gives $\nabla\varphi = 2r^{-1}|\psi|^{(2/r)-1}(\text{sgn }\psi)\nabla\psi$ so that Hölder's inequality applied to (5.35) results in

$$||q_-^{1/2}|\psi|^{2/r}||_{L^2(B)} \le c''\{||\nabla\psi||_{L^2(B)}||\psi||^{(2-r)/r} + ||\psi||^{2/r}_{L^2(B)}\} \qquad (5.36)$$

A straightforward limiting argument shows that (5.36) is valid for $\psi \in H_0^1(B)$. We shall substitute $\psi = \zeta^{r/2}|u|^{p_0/2}$. (Note that $\zeta^{r/2} = \zeta_1^r \in C_0^1(B)$.) Thus

$$||q_-^{1/2}\zeta|u|^{(2/r)(p_0/2)}||_{L^2(B)} \le c''\{||\nabla(\zeta^{r/2}|u|^{p_0/2})||_{L^2(B)}||\zeta^{r/2}|u|^{p_0/r}||^{(2-r)/r}_{L^2(B)}$$
$$+ ||\zeta^{r/2}|u|^{p_0/2}||^{2/r}_{L^2(B)}\},$$

so that $\int_B \zeta^2 q_-|u|^p dx < \infty$ if $p \le (2/r)p_0$.

Thus we see that the right side of (5.31) is finite if $p = \eta p_0$ with $\eta = \min\{2/r, (2^*+2)/4\}$ and if we choose $\zeta = \zeta_1^2$, ζ_1 any non-negative function in $C_0^\infty(B)$. It thus follows from (5.31) that for this p

$$\int_B \zeta^2|\nabla_A|u||^2|u|^{p-2}dx < \infty.$$

Using our freedom to vary ζ it follows that $(\partial_i|u|)|u|^{(p/2)-1} \in L^2_{loc}(B)$. From (5.34) it follows that $|u|^{p/2} \in L^2_{loc}(B)$ and thus Corollary 5.6 allows us to write

$$\partial_i|u|^{p/2} = (p/2)|u|^{(p/2)-1}\partial_i|u| \qquad (5.37)$$

and conclude that $\nabla|u|^{p/2} \in L^2_{loc}(B)$. Hence $|u|^{p/2} \in H^1_{loc}(B)$. By induction it follows that $|u|^{p/2} \in H^1_{loc}(B)$ for all $p \ge 2$.

To find explicit bounds we go back to (5.31) which we rewrite in the form

$$\frac{4(p-1)}{p^2}\int_B \zeta^2|\nabla_A|u|^{p/2}|^2dx \le 2\int_B \zeta|\nabla_A|u|\cdot\nabla_A\zeta||u|^{p-1}dx$$
$$+\int_B q_-\zeta^2|u|^p\,dx \tag{5.38}$$

where we have used (5.37) to write $|\nabla_A|u|^{p/2}|^2 = (p^2/4)|u|^{p-2}|\nabla_A|u||^2$. We no longer make the assumption that $\zeta = \zeta_1^2$ but continue to assume $\zeta \in C_0^\infty(B)$ and non-negative. Using (5.32) with $p_0 = p$ to bound the first term on the right side of (5.38) results in

$$\int_B \zeta^2|\nabla_A|u|^{p/2}|^2dx \le (\frac{p}{p-1})(\int_B \zeta^2|\nabla_A|u|^{p/2}|^2dx)^{1/2}(\int_B|\nabla_A\zeta|^2|u|^p\,dx)^{1/2}$$
$$+\frac{p^2}{4(p-1)}\int_B q_-\zeta^2|u|^p\,dx. \tag{5.39}$$

This inequality is of the form

$$A^2 \le (\frac{p}{p-1})AB + \frac{p^2}{4(p-1)}C.$$

Since $AB \le A^2/4 + B^2$ we have

$$A^2[1-\frac{p}{4(p-1)}] \le \frac{p}{p-1}B^2 + \frac{p^2}{4(p-1)}C.$$

Since $p \ge 2$, $p/(p-1) \le 2$, and thus (5.39) implies

$$\int_B \zeta^2|\nabla_A|u|^{p/2}|^2dx \le 4\int_B |u|^p|\nabla_A\zeta|^2dx + p\int_B q_-\zeta^2|u|^p\,dx \tag{5.40}$$

and upon taking square roots

$$||\zeta|\nabla_A|u|^{p/2}|\,||_{L^2(B)} \le 2||\,|u|^{p/2}|\nabla_A\zeta|\,||_{L^2(B)}$$
$$+\sqrt{p}\,||q_-^{1/2}\zeta|u|^{p/2}||_{L^2(B)}. \tag{5.41}$$

According to (5.1) and (5.25) there is a number $\vartheta \in (0,1)$ and constants c, c_1 such that for any $\varepsilon > 0$ and $\varphi \in C_0^\infty(B)$:

$$||q_-^{1/2}\varphi||_{L^2(B)} \le c_1\{\varepsilon||\nabla\varphi||_{L^2(B)} + c(\varepsilon^{-\vartheta/(1-\vartheta)}+\varepsilon)||\varphi||_{L^2(B)}\} \tag{5.42}$$

Since (5.42) clearly extends to $\varphi \in H_0^1(B)$ we substitue $\varphi = \zeta|u|^{p/2}$. Putting the resulting inequality into (5.41) and using $|\nabla\varphi| \le c_2^{-1/2}|\nabla_A\varphi|$ results in

$$||\zeta|\nabla_A|u|^{p/2}|\,||_{L^2(B)} \le 2||\,|u|^{p/2}|\nabla_A\zeta|\,||_{L^2(B)}$$

$$+\sqrt{p}\ c_3\{\varepsilon\|\nabla_A(\zeta|u|^{p/2})\|_{L^2(B)}+(\varepsilon^{-\vartheta(1-\vartheta)}+\varepsilon)\|\zeta|u|^{p/2}\|_{L^2(B)}\} \quad (5.43)$$

where we have abbreviated $\||\nabla_A v|\|_{L^2(B)}=\|\nabla_A v\|_{L^2(B)}$ and $|\nabla_A v|=(|\nabla_A v|^2)^{1/2}$. Using

$$\|\nabla_A(\zeta|u|^{p/2})\|_{L^2(B)}\le\||\nabla_A\zeta||u|^{p/2}\|_{L^2(B)}+\|\zeta|\nabla_A|u|^{p/2}|\|_{L^2(B)}$$

and choosing ε such that $c_3\sqrt{p}\,\varepsilon=1/2$, we find

$$\|\nabla_A(\zeta|u|^{p/2})\|_{L^2(B)}\le 6\||u|^{p/2}|\nabla_A\zeta|\|_{L^2(B)}$$

$$+\ 2c_3p^{1/2}(\varepsilon^{-\vartheta/(1-\vartheta)}+\varepsilon)\|\zeta|u|^{p/2}\|_{L^2(B)}$$

$$\le\{6+2c_3p^{1/2}(\varepsilon^{-\vartheta(1-\vartheta)}+\varepsilon)\}\||u|^{p/2}(|\nabla_A\zeta|+\zeta)\|_{L^2(B)}$$

$$\le c_4p^{\alpha}\||u|^{p/2}(|\nabla_A\zeta|+\zeta)\|_{L^2(B)} \quad (5.44)$$

where $\alpha=[2(1-\vartheta)]^{-1}$. We put this in a form which is useful for deriving L^∞ estimates by using Sobolev's inequality, Eqn. (5.33):

$$\|\zeta|u|^{p/2}\|_{L^{2^*}(B)}\le c_5p^{\alpha}\||u|^{p/2}(|\nabla_A\zeta|+\zeta)\|_{L^2(B)}. \quad (5.45)$$

Let $a\in(0,1)$ and choose $\varepsilon>0$ so that with $\varepsilon_i=\varepsilon/i^2$, $1-a=\sum_{i=1}^{\infty}\varepsilon_i$. Defining $a_i=1-\sum_{m=1}^{i}\varepsilon_m$ we note that $\{a_i\}_{i=1}^{\infty}$ is a decreasing sequence converging to a. Let

$$\zeta_i(x)=\begin{cases}1 & \text{if } |x|<a_{i+1}\\ (a_i-a_{i+1})^{-1}(a_i-|x|) & \text{if } a_{i+1}\le|x|\le a_i\\ 0 & \text{if } a_i<|x|.\end{cases}$$

Since (5.45) clearly extends to $\zeta=\zeta_i$ and since $|\nabla_A\zeta_i|\le(Const.)(i+1)^2$ we thus have

$$\||u|^{p/2}\|_{L^{2^*}(B_{a_{i+1}})}\le c_6p^{\alpha}(i+1)^2\||u|^{p/2}\|_{L^2(B_{a_i})}, \quad (5.46)$$

B_{a_i} denoting the ball $B(0;a_i)$. We take $p=p_i=2(2^*/2)^{i-1}$. Rearranging (5.46) we find

$$\|u\|_{L^{p_{i+1}}(B_{a_{i+1}})}\le[c_6p_i^{\alpha}(i+1)^2]^{2^*/p_{i+1}}\|u\|_{L^{p_i}(B_{a_i})}. \quad (5.47)$$

Because p_i increases exponentially with i it is easy to see that for some constant D,

$$\prod_{i=1}^{\ell}[c_6 p_i^{a}(i+1)^2]^{2^*/p_{i+1}} \le D$$

for all ℓ so that (5.47) yields the inequality

$$||u||_{L^{p_{i+1}}(B_a)} \le ||u||_{L^{p_{i+1}}(B_{a_{i+1}})} \le D||u||_{L^2(B_{a_1})}$$

which holds for $i = 1,2,\ldots$. Letting $i \to \infty$ it follows that

$$\operatorname*{ess\,sup}_{x\in B_a} |u(x)| = \lim_{i\to\infty}||u||_{L^{p_i}(B_a)} \le D||u||_{L^2(B)}$$

This yields (5.5) and completes the proof of the theorem. ▪

Appendix 1 Approximation of Metrics and Completeness

Lemma A1.1. *Let $g_{ij}(x)$, $1 \le i,j \le n$, be real valued continuous functions on an open subset Ω of $\mathbb{R}^n$ such that $[g_{ij}(x)] = G(x)$ is a positive definite matrix for every $x \in \Omega$ Then for every $\delta > 0$ there exists a positive definite matrix $[h_{ij}(x)] = H(x)$ with real $h_{ij} \in C^\infty(\Omega)$ such that*

$$(1-\delta)\ G(x) \le H(x) \le (1+\delta)\ G(x) \tag{A1 1}$$

for every $x \in \Omega$

We recall that for $n \times n$ real symmetric matrices A and B, $A \le B$ means $\langle A\xi,\xi\rangle \le \langle B\xi,\xi\rangle$ for every $\xi \in \mathbb{R}^n$ We also let

$$||A|| = \sup\{|A\xi| \quad \xi \in \mathbb{R}^n\ ,\ |\xi| \le 1\}$$

Proof of Lemma: Choose a partition of unity $\{\varphi_j \quad j = 1,2, \quad \}$ for Ω satisfying

(i) $\varphi_j \in C_0^\infty(\Omega)$

(ii) For any compact set K, supp $\varphi_j \cap K$ is empty for all but finitely many j

Let $\zeta \in C_0^\infty(\mathbb{R}^n)$ with $\int \zeta(x)dx = 1$ and supp $\zeta \subset B(0,1)$ and let $\zeta_\varepsilon(x) = \varepsilon^{-n}\zeta(x/\varepsilon)$ Let K_j be a compact subset of Ω with int $K_j \supset$ supp φ_j Let $\gamma_j = 2^{-j}\delta \inf_{x \in K_j}(||G(x)^{-1}||)^{-1}$ and choose $0 < \varepsilon_j < 1/j$ so that

(a) supp $(\zeta_{\varepsilon_j} * (\varphi_j G)) \subset K_j$ (Here $*$ denotes the convolution operation on functions on $\mathbb{R}^n$)

(b) $||(\zeta_{\varepsilon_j} * (\varphi_j G))(x) - \varphi_j(x) G(x)|| \le \gamma_j$

for all $x \in \Omega$ Then

$$-\gamma_j I \le \zeta_{\varepsilon_j} * (\varphi_j G) - \varphi_j G \le \gamma_j I \tag{A1 2}$$

Since $G(x) \ge I\,||G(x)^{-1}||^{-1}$ it follows from the definition of γ_j that

$\gamma_j I \leq 2^{-j}\delta G(x)$ for $x \in K_j$. Thus by (A1 2) and (a),

$$\varphi_j(x)G(x) - 2^{-j}\delta G(x) \leq (\zeta_{\varepsilon_j} *(\varphi_j G))(x) \leq \varphi_j(x)G(x) + 2^{-j}\delta G(x) \quad \text{(A1 3)}$$

Let

$$H(x) = \sum_{j=1}^{\infty} (\zeta_{\varepsilon_j} *(\varphi_j G))(x) \quad \text{(A1 4)}$$

We show that H is well defined and satisfies the desired inequalities Suppose Ω_0 is an open subset of Ω with a compact closure in Ω Then for large enough j $\text{dist}(\overline{\Omega}_0, \text{supp}\, \varphi_j) \geq \alpha > 0$ for some α because from (ii) it follows that if K is a compact subset of Ω with $\text{int}\, K \supset \overline{\Omega}_0$, then for large enough j $\text{supp}\, \varphi_j \cap K$ is empty For these j $\text{dist}(\overline{\Omega}_0, \text{supp}\, \varphi_j) \geq \text{dist}((\text{int}\, K)^c, \overline{\Omega}_0) > 0$ Thus because $\varepsilon_j < 1/j$, for large enough j $\text{dist}(\overline{\Omega}_0, \text{supp}\, (\zeta_{\varepsilon_j} *(\varphi_j G))) > 0$ This implies that for x in Ω_0 all terms in the sum (A1 4) for j large enough vanish identically Hence H is well defined and belongs to $C^\infty(\Omega)$ Finally, summing the inequalities (A1 3) over j, using $\Sigma\varphi_j = 1$, gives the desired inequalities ▪

Lemma A1.2 *Let Ω be an open connected subset of $\mathbf{R}^n$ and $a_{ij}(x)$ continuous functions on Ω such that $[a_{ij}(x)]$ is positive definite for every x Let ρ_λ be as given by (1 9) where $\lambda(x) > 0$ is continuous and let*

$$\rho_\lambda(x, \{\infty\}) = \sup\{\rho_\lambda(x, \Omega \setminus K) \quad K \text{ compact } \}$$

Then (Ω, ρ_λ) is a complete metric space iff $\rho_\lambda(x, \{\infty\}) = \infty$

Proof Suppose $\rho_\lambda(x, \{\infty\}) = \infty$ and $\{x_j\}$ is a ρ_λ-Cauchy sequence Then, since

$$|\rho_\lambda(x_0, x_j) - \rho_\lambda(x_0, x_m)| \leq \rho_\lambda(x_j, x_m)$$

$\rho_\lambda(x_0, x_j)$ is a Cauchy sequence of numbers which therefore converges and thus is bounded Since $\rho_\lambda(x, \{\infty\}) = \infty$ we can find a compact set K such that $\{x_j\} \subset K$ It is easy to see that ρ_λ generates the usual topology on Ω Thus (K, ρ_λ) is a compact metric space and hence complete Therefore $\{x_j\}$ converges

Suppose, on the other hand, that (Ω, ρ_λ) is complete Assume in addition that the matrix $\lambda(x)[a_{ij}(x)]$ has entries in $C^\infty(\Omega)$ Then the infimum over absolutely continuous paths in the definition of ρ_λ can be replaced by an infimum over smooth paths and ρ_λ is the metric treated in standard differential geometry text books [22, 13] We sketch the rest of the proof using ideas from differential geometry By the Hopf-Rinow theorem [18] metric space completeness is equivalent to geodesic completeness This implies that given $x \in \Omega$ and ξ in the tangent space T_x at x, there exists a geodesic $\gamma(t)$ defined for all t such that $\gamma(0) = x$, $\dot\gamma(0) = \xi$ For $x \in \Omega$ consider the map Exp_x $T_x \to \Omega$ which maps $\xi \in T_x$ to $\gamma(1)$ where γ is the geodesic such that $\gamma(0) = x$ and $\dot\gamma(0) = \xi$ Geodesic completeness implies Exp_x is defined on all of T_x Since Exp_x is continuous the sets $\mathrm{Exp}_x(\overline{B(0,N)})$ $N = 1,2,3,$ where

$$B(0,N) = \{\xi \in T_x \quad |\xi|_x = \lambda(x)^{1/2}(\Sigma a_{ij}(x)\xi_i\xi_j)^{1/2} < N\}$$

are compact and

$$\rho_\lambda(x, \Omega \setminus \mathrm{Exp}_x(\overline{B(0,N)})) \geq N$$

This shows $\rho_\lambda(x,\{\infty\}) = \infty$

In the case where the matrix $\lambda(x)[a_{ij}(x)]$ does not necessarily have smooth entries we apply the approximation Lemma A1 1 to find a matrix $H(x)$ with smooth entries, such that for some $\delta \in (0,1)$

$$(1-\delta)\lambda(x)[a_{ij}(x)] \leq H(x) \leq (1+\delta)\lambda(x)[a_{ij}(x)]$$

Then if $\tilde\rho$ denotes the distance function defined by H we have, for $x,y \in \Omega$

$$(1-\delta)\rho_\lambda(x,y) \leq \tilde\rho(x,y) \leq (1+\delta)\rho_\lambda(x,y)$$

So if Ω is complete in the metric ρ_λ and $\{x_j\}$ is a $\tilde\rho$-Cauchy sequence

$$\rho_\lambda(x_j,x_m) \leq (1-\delta)^{-1}\tilde\rho(x_j,x_m)$$

Thus the sequence is also ρ_λ-Cauchy and hence converges This shows Ω is $\tilde\rho$-complete By the previous argument $\tilde\rho(x,\{\infty\}) = \infty$ so that $\rho_\lambda(x,\{\infty\}) \geq (1+\delta)^{-1}\tilde\rho(x,\{\infty\}) = \infty$ which completes the proof ▪

Appendix 2 Proof of Lemma 1.2

The fact that K_* exists and satisfies (2) and (3) follows immediately from the definitions while (4) is a consequence of the fact that for fixed x , $K_*(x,\cdot)$ is the supremum of linear functions. To see continuity let $G \subset \Omega$ be compact and define, for $x \in G$ and η, $\xi \in S^{n-1}$,

$$f_\eta(x,\xi) = \langle \xi,\eta \rangle / K(x,\eta)$$

so that $K_*(x,\xi) = \sup\{f_\eta(x,\xi) : \eta \in S^{n-1}\}$. Given $x \in G$ and $\xi \in S^{n-1}$ choose $\eta_0 \in S^{n-1}$ so that $K_*(x,\xi) = f_{\eta_0}(x,\xi)$. Then for $x' \in G$ and $\xi' \in S^{n-1}$

$$K_*(x',\xi') \geq f_{\eta_0}(x',\xi'),$$

so that

$$\begin{aligned} K_*(x,\xi) &= f_{\eta_0}(x,\xi) = f_{\eta_0}(x',\xi') + (f_{\eta_0}(x,\xi) - f_{\eta_0}(x',\xi')) \\ &\leq K_*(x',\xi') + \sup_{\eta \in S^{n-1}} |f_\eta(x,\xi) - f_\eta(x',\xi')|. \end{aligned}$$

Reversing the roles of (x,ξ) and (x',ξ') we thus have

$$|K_*(x,\xi) - K_*(x',\xi')| \leq \sup_{\eta \in S^{n-1}} |f_\eta(x,\xi) - f_\eta(x',\xi')|.$$

Since $f_\eta(x,\xi)$ is uniformly continuous for η, $\xi \in S^{n-1}$ and $x \in G$, the continuity of K_* follows.

The fact that $K_{**} = K$ follows from the bipolar theorem [33; Theorem V. 28]. We give a direct proof. Since the variable x is irrelevant we suppress it.

Since $K_*(\xi) = \sup\{\langle \xi,\eta \rangle / K(\eta) : \eta \neq 0\}$ we have

$$K_*(\xi)K(\eta) \geq \langle \xi,\eta \rangle \text{ for } all \ \ \xi, \eta .$$

Thus

$$K(\eta) \geq \sup_{\xi \neq 0} \langle \xi,\eta \rangle / K_*(\xi) = K_{**}(\eta).$$

We now prove the reverse inequality. Let

$$C = \{\eta : \ K(\eta) \leq 1\}.$$

Note that C is a compact convex set. We have

$$\begin{aligned} K_*(\xi) &= \sup\{<\xi,\eta>/K(\eta) : \eta \neq 0\} \\ &= \sup\{<\xi,\eta/K(\eta)> : \eta \neq 0\} \\ &= \sup\{<\xi,\omega> : K(\omega) = 1\} \\ &= \sup\{<\xi,\omega> : \omega \in C\}. \end{aligned}$$

The last equality is a consequence of the fact that the supremum is taken on for some ω with $K(\omega) = 1$.

Since K_{**} enjoys the same properties as K we have

$$K_{**}(\xi) = \sup\{<\xi,\omega> : \omega \in C^0\}$$

where C^0, the polar of C, is given by

$$\begin{aligned} C^0 &= \{\omega : K_*(\omega) \leq 1\} \\ &= \{\omega : <\omega,\eta> \leq 1 \quad \text{for} \quad \eta \in C\}. \end{aligned} \tag{A2.1}$$

Now suppose there is an η_0 with

$$K(\eta_0) > K_{**}(\eta_0).$$

Then by multiplying η_0 by a positive constant we can assume that

$$K(\eta_0) > 1, \quad K_{**}(\eta_0) < 1. \tag{A2.2}$$

Thus $\eta_0 \notin C$ and since C is a compact convex set we can find a linear functional $\ell : \mathbb{R}^n \to \mathbb{R}$ with $\ell(\eta_0) = a$ and $\ell(\xi) < a$ for $\xi \in C$. Since $0 \in C$, $0 = \ell(0) < a$. Therefore if τ_0 is the vector such that $<\tau_0,\cdot> = a^{-1}\ell(\cdot)$ we have $<\tau_0,\eta_0> = 1$ and $<\tau_0,\xi> < 1$ for all $\xi \in C$. So $\tau_0 \in C^0$ by (A2.1). Hence

$$\begin{aligned} K_{**}(\eta_0) &= \sup\{<\eta_0,\tau> : \tau \in C^0\} \\ &\geq <\eta_0,\tau_0> = 1. \end{aligned}$$

This contradicts (A2.2). Therefore $K_{**}(\eta) \geq K(\eta)$ for every η which completes the proof. ▪

Appendix 3 Proof of Lemma 2 2

Let us prove that $\Lambda_R(x, A+q)$ is continuous in x and R Fixing $P = A+q$ we simplify notation and write $\Lambda_R(x)$ for $\Lambda_R(x, A+q)$ We consider $\Lambda_R(x)$ as a function of (x, R) in $\mathbf{R}^n \times \mathbf{R}_+$ with the usual Euclidean metric We start by showing that $\Lambda_R(x)$ is an everywhere finite upper semicontinuous function

Let Ω be a bounded open set in $\mathbf{R}^n$ Since $q_- \restriction \Omega$ belongs to $M(\Omega)$ and since $[a^{ij}(x)] \geq \delta I$ for every $x \in \Omega$ for some $\delta > 0$, it follows from Lemma 0 3 (applied with $\varepsilon = \min(\frac{1}{2}, \frac{\delta}{2})$) that there exists a constant C_Ω, depending on Ω and A, such that

$$\int_\Omega q_- |\varphi|^2 dx \leq \frac{1}{2} \int_\Omega |\nabla_A \varphi|^2 dx + C_\Omega \int_\Omega |\varphi|^2 dx \tag{A3 1}$$

for every $\varphi \in C_0^\infty(\Omega)$ Using (A3 1) it follows that

$$\begin{aligned}(P\varphi, \varphi) &= \int_\Omega (|\nabla_A \varphi|^2 + q_+ |\varphi|^2 - q_- |\varphi|^2) dx \\ &\geq \frac{1}{2} \int_\Omega (|\nabla_A \varphi|^2 + q_+ |\varphi|^2) dx - C_\Omega \int_\Omega |\varphi|^2 dx \end{aligned} \tag{A3 2}$$

for every $\varphi \in C_0^\infty(\Omega)$ This inequality implies in particular that $\Lambda_R(x) \geq - C_\Omega$ whenever $B(x, R) \subset \Omega$ Thus, since Ω is an arbitrary bounded open set, $\Lambda_R(x)$ is everywhere finite Now let (x^0, R_0) be a fixed point in $\mathbf{R}^n \times \mathbf{R}_+$ and let $\{(x^j, R_j)\}$, $j = 1, 2,$, be a sequence of points in $\mathbf{R}^n \times \mathbf{R}_+$ such that $(x^j, R_j) \to (x^0, R_0)$ as $j \to \infty$ Pick a number $\varepsilon > 0$ and choose a function $\psi \in C_0^\infty(B(x^0, R_0))$, such that $||\psi|| = ||\psi||_{L^2} = 1$, and such that

$$(P\psi, \psi) \leq \Lambda_{R_0}(x^0) + \varepsilon$$

With ψ defined as zero in $\mathbf{R}^n \setminus B(x^0, R_0)$, it is clear that supp $\psi \subset B(x^j, R_j)$ for j large enough Hence, for large j we have

$$\begin{aligned}\Lambda_{R_j}(x^j) &= \inf \{(P\varphi, \varphi) \quad \varphi \in C_0^\infty(B(x^j, R_j)), \ ||\varphi|| = 1\} \\ &\leq (P\psi, \psi) \leq \Lambda_{R_0}(x^0) + \varepsilon\end{aligned}$$

Letting $j \to \infty$ in the last inequality and then letting $\varepsilon \to 0$, we get

$$\limsup_{j\to\infty} \Lambda_{R_j}(x^j) \le \Lambda_{R_0}(x^0). \tag{A3.3}$$

This proves that $\Lambda_R(x)$ is an upper semicontinuous function.

Next we show that $\Lambda_R(x)$ is a lower semicontinuous function which together with the preceding result will prove that $\Lambda_R(x)$ is continuous in (x,R). To this end we introduce some classes of functions.

Let Ω be a bounded open set in $\mathbb{R}^n$. We denote by $H^1_{oo}(\Omega)$ the completion of $C_0^\infty(\Omega)$ in the $H^1(\Omega)$ norm:

$$\|u\|^2_{H^1(\Omega)} = \int_\Omega (\sum_{i=1}^n |\partial_i u|^2 + |u|^2)dx.$$

We also introduce the space of functions

$$H^1_{oq_+}(\Omega) = \{u : H^1_{oo}(\Omega)\, ,\, q_+^{1/2} u \in L^2(\Omega)\} \tag{A3.4}$$

where q_+ is the positive part of the function q in the lemma. $H^1_{oq_+}(\Omega)$ is a Hilbert space under the norm $|||\cdot|||_\Omega$ defined by

$$|||u|||^2_\Omega = \|u\|^2_{H^1(\Omega)} + \|q_+^{1/2} u\|^2_{L^2(\Omega)}. \tag{A3.5}$$

We shall need the fact that $C_0^\infty(\Omega)$ is dense in $H^1_{oq_+}(\Omega)$. To prove this result suppose that $u \in H^1_{oq_+}(\Omega)$ is orthogonal to $C_0^\infty(\Omega)$ in $H^1_{oq_+}(\Omega)$, that is

$$\int_\Omega (\nabla u \cdot \nabla \varphi + (1+q_+)u\varphi)dx = 0$$

for every $\varphi \in C_0^\infty(\Omega)$. This means that $-\Delta u = -(1+q_+)u$ weakly. Since $(1+q_+)u \in L^1_{loc}(\Omega)$ we are in a position to apply Lemma 5.4. It follows from the lemma that $|u| \in H^1_{loc}(\Omega)$ and that $-\Delta|u| \le -\mathrm{Re}((1+q_+)u(\mathrm{sgn}\,\bar{u})) \le -|u|$ in the sense that

$$\int_\Omega (\nabla|u|\cdot\nabla\varphi + |u|\varphi)dx \le 0 \tag{A3.6}$$

for every $\varphi \in C_0^\infty(\Omega)$, $\varphi \ge 0$. Since $u \in H^1_{oo}(\Omega)$ we also have that $|u| \in H^1_{oo}(\Omega)$. Postponing the proof of this result, let $\{\psi_j\}$ be a sequence of real functions in $C_0^\infty(\Omega)$ such that $\psi_j \to |u|$ in $H^1_{oo}(\Omega)$. Choosing if necessary a subsequence we may assume that $\psi_j(x) \to |u(x)|$ almost everywhere. Set

$\varphi_j = (\psi_j^2+j^{-2})^{1/2} - j^{-1}$, $j = 1,2,$ Then $\varphi_j \in C_0^\infty(\Omega)$, $\varphi_j \geq 0$ and it is easily checked that $\varphi_j \to |u|$ in $H_{oo}^1(\Omega)$ (here one uses the fact that $\partial_i |u|$ vanishes almost everywhere on the set $S_0 = \{x \;\; u(x) = 0\}$, see Lemma 5 5) Applying (A3 6) with $\varphi = \varphi_j$ and letting $j \to \infty$ it folows that $\int_\Omega (|\nabla u|^2 + |u|^2)dx \leq 0$ which implies that $u = 0$

To complete the proof that $C_0^\infty(\Omega)$ is dense in $H_{oq_+}^1(\Omega)$ we therefore need only to show that if $u \in H_{oo}^1(\Omega$ then also $|u| \in H_{oo}^1(\Omega)$ Suppose first that $u \in C_0^\infty(\Omega)$ Set $\varphi_j(x) = (|u|^2+j^{-2})^{1/2} - j^{-1}$, $j = 1,2,$ Then $\varphi_j \in C_0^\infty(\Omega)$ and it is clear that $\varphi_j(x) \to |u(x)|$ as $j \to \infty$ uniformly in Ω We have

$$\partial_i \varphi_j = \mathrm{Re}(u(|u|^2+j^{-2})^{-1/2}\partial_i \bar{u}) \tag{A3 7}$$

which shows that the sequence $\{\varphi_j\}$ is bounded in $H_{oo}^1(\Omega)$ These properties imply that there exists a function $\tilde{u} \in H_{oo}^1(\Omega)$ such that $\varphi_j \to \tilde{u}$ weakly in $H_{oo}^1(\Omega)$, and that $\tilde{u}(x)$ coincides with $|u(x)|$ for almost all x This proves that $|u| \in H_{oo}^1(\Omega)$ Using (A3 7) we also see that

$$\| |u| \|_{H^1(\Omega)} \leq \lim_{j\to\infty} \sup \|\varphi_j\|_{H^1(\Omega)} \leq \|u\|_{H^1(\Omega)}$$

Now consider the general case Let $u \in H_{oo}^1(\Omega)$ and let $\{\varphi_j\}$ be a sequence of functions in $C_0^\infty(\Omega)$ such that $\varphi_j \to u$ in $H_{oo}^1(\Omega)$ From the result just proved it follows that $|\varphi_j| \in H_{oo}^1(\Omega)$ and that $\| |\varphi_j| \|_{H^1(\Omega)} \leq \|\varphi_j\|_{H^1(\Omega)}$ which implies that $\{|\varphi_j|\}$ is a bounded sequence in $H_{oo}^1(\Omega)$ Since $|\varphi_j| \to |u|$ in $L^2(\Omega)$, it follows as above that $|u| \in H_{oo}^1(\Omega)$

We proceed now with the proof that $\Lambda_R(x)$ is a lower semicontinuous function in x and R We fix a point (x^0,R_0) in $\mathbb{R}^n \times \mathbb{R}_+$ and pick a bounded open set Ω such that $\Omega \supset \overline{B(x^0,R_0)}$ By adding to P a sufficiently large constant and denoting the resulting operator again by P, we may assume (in view of the estimate $[a^{ij}(x)] \geq \delta I$, which holds for every $x \in \Omega$ and some $\delta \in (0,1)$ and the consequent estimate (A3 2)) that

$$(P\varphi,\varphi) \geq \frac{1}{2}\int_\Omega (|\nabla_A \varphi|^2 + |\varphi|^2 + q_+|\varphi|^2)dx \geq \frac{\delta}{2}|||\varphi|||_\Omega^2 \tag{A3 8}$$

for every $\varphi \in C_0^\infty(\Omega)$ Here $|||\;\;|||_\Omega$ is the norm in $H_{oq_+}^1(\Omega)$ introduced before

Also, since the $a^{ij}(x)$ are bounded in Ω it follows that

$$(P\varphi,\varphi) \leq c\,|||\varphi|||_{\Omega}^2 \tag{A3 9}$$

for every $\varphi \in C_0^\infty(\Omega)$ and for some constant c Let us write

$$(Pu,u)_\Omega = \int_\Omega (|\nabla_A u|^2 + q\,|u|^2)dx \tag{A3 10}$$

for any function $u \in H^1(\Omega)$ such that $q_+^{1/2} u \in L^2(\Omega)$ In view of the bounds (A3 8) and (A3 9) it follows (since $C_0^\infty(\Omega)$ is dense in $H^1_{oq_+}(\Omega)$) that $(Pu,u)_\Omega^{1/2}$ is an equivalent norm in the Hilbert space $H^1_{oq_+}(\Omega)$

We now consider a sequence of points $\{(x^j,R_j)\}$ in $\mathbf{R}^n \times \mathbf{R}_+$ such that $(x^j,R_j) \to (x^0,R_0)$ as $j \to \infty$ where (x^0,R_0) is the point we have fixed before Recalling that $\overline{B(x^0,R_0)} \subset \Omega$ we shall assume with no loss of generality that $\overline{B(x^j,R_j)} \subset \Omega$ for all j We now choose a sequence of functions $\varphi_j \in C_0^\infty(B(x^j,R_j))$ such that

$$(P\varphi_j,\varphi_j) \leq \Lambda_{R_j}(x^j) + \frac{1}{j} \quad \text{and} \quad ||\varphi_j|| = 1 \tag{A3 11}$$

for $j = 1,2,$ We consider the φ_j as elements of $H^1_{oq_+}(\Omega)$ (φ_j being defined as zero in $\Omega \setminus B(x^j,R_j)$) Since $\{\Lambda_{R_j}(x^j)\}$ is a bounded sequence (by (A3 3)), it follows that the sequence of norms $(P\varphi_j,\varphi_j)_\Omega^{1/2}$ is bounded Thus we conclude that $\{\varphi_j\}$ is a bounded sequence in $H^1_{oq_+}(\Omega)$ We now pick a subsequence $\{\varphi_{j_k}\}$ which converges weakly in $H^1_{oq_+}(\Omega)$ Changing notation and denoting this subsequence again by $\{\varphi_j\}$, we thus may assume that there exists $u \in H^1_{oq_+}(\Omega)$ such that

$$\text{weak} \lim_{j\to\infty} \varphi_j = u \quad \text{in} \quad H^1_{oq_+}(\Omega) \tag{A3 12}$$

Applying Rellich's compactness theorem [3, Theorem 3 8] it follows from (A3 12) that

$$\varphi_j \to u \quad \text{strongly in} \quad L^2(\Omega) \tag{A3 13}$$

Since u is a weak limit of $\{\varphi_j\}$ in $H^1_{oq_+}(\Omega)$, with the norm of φ_j given by $(P\varphi_j,\varphi_j)_\Omega^{1/2}$, we conclude that

$$(Pu,u)_\Omega \le \liminf_{j\to\infty} (P\varphi_j,\varphi_j) \le \liminf_{j\to\infty} \Lambda_{R_j}(x^j) \tag{A3 14}$$

where the last inequality follows from (A3 11) Observe also that from (A3 11) and (A3 13) it follows that $\|u\| = 1$ Denote by u_0 the restriction of u to $B(x^0,R_0)$ and note that $u = 0$ a e in $\Omega\setminus B(x^0,R_0)$ (in view of (A3 13) since φ_j is supported in $B(x^j,R_j)$) Hence using (A3 14) we have

$$(Pu_0,u_0)_{B(x^0,R_0)} = (Pu,u)_\Omega \le \liminf_{j\to\infty} \Lambda_{R_j}(x^j) \tag{A3 15}$$

We now claim that $u_0 \in H^1_{oo}(B(x^0,R_0))$ which implies (since $q_+^{1/2}u_0 \in L^2(B(x^0,R_0))$ that $u_0 \in H^1_{oq_+}(B(x^0,R_0))$ Accepting this result for a moment we proceed with the proof and show that $\Lambda_R(x)$ is lower semicontinuous at (x^0,R_0) Indeed, since as we have shown $C_0^\infty(B(x^0,R_0))$ is dense in $H^1_{oq_+}(B(x^0,R_0))$, it follows that there exists a sequence of functions $\psi_j \in C_0^\infty(B(x^0,R_0))$ such that $\psi_j \to u_0$ in $H^1_{oq_+}(B(x^0,R_0))$ We also have that $\|u_0\|_{L^2(B(x^0,R_0))} = \|u\|_{L^2(\Omega)} = 1$, and thus we may assume that $\|\psi_j\|_{L^2(B(x^0,R_0)} = 1$ Since $\Lambda_{R_0}(x^0) \le (P\psi_j,\psi_j)_{B(x^0,R_0)}$ for $j = 1,2,$, we obtain upon passage to the limit that

$$\Lambda_{R_0}(x^0) \le (Pu_0,u_0)_{B(x^0\ R_0)}$$

which when combined with (A3 15) shows that

$$\Lambda_{R_0}(x^0) \le \liminf_{j\to\infty} \Lambda_{R_j}(x^j)$$

This proves that $\Lambda_R(x)$ is a lower semicontinuous function and thus establishes (since $\Lambda_R(x)$ is also upper semicontinuous) that $\Lambda_R(x)$ is continuous

We still must show that $u_0 \in H^1_{oo}(B(x^0,R_0)$ To this end introduce the affine maps T_j $\mathbf{R}^n \to \mathbf{R}^n$ which map $B(x^0,R_0)$ onto $B(x^j,R_j)$ $T_jx = \vartheta_j(x-x^0) + x^j$ with $\vartheta_j = R_j/R_0$ Consider the sequence of functions $\psi_j = \varphi_j \circ T_j$, $j = 1,2,$, where $\varphi_j \in C_0^\infty(B(x^j,R_j))$ is the sequence of functions introduced above Then $\psi_j \in C_0^\infty(B(x^0,R_0))$ Consider ψ_j and φ_j as elements of $H^1_{oo}(\Omega)$ Since $x^j \to x_0$, $\vartheta_j \to 1$ and since $\{\varphi_j\}$ is a bounded sequence in $H^1_{oo}(\Omega)$ it follows readily that $\psi_j - \varphi_j \to 0$ weakly in $H^1_{oo}(\Omega)$, so that in view of (A3 12) it follows that $\psi_j \to u$ weakly in $H^1_{oo}(\Omega)$ Since the ψ_j are supported in the ball $B(x^0;R_0)$ it follows further that when restricted to this ball $\psi_j \to u_0$ weakly in $H^1(B(x^0;R_0))$. This implies that $u_0 \in H_{oo}(B(x^0;R_0))$ as claimed. The main part of the lemma is thus proved.

We shall now prove the statement that if $A_m(x,\partial) = -\sum_{i,j}\partial_j a_m^{ij}\partial_i$, $m = 1,2,\ldots,$
is a sequence of operators satisfying the same conditions as A and if $a_m^{ij} \to a^{ij}$ as $m \to \infty$ uniformly on compact sets, then $\Lambda_R(x;A_m+q) \to \Lambda_R(x;A+q)$ uniformly in x on compact sets. Thus let K be a compact set in $\mathbb{R}^n$ and let $R > 0$ be fixed. Pick a bounded open set Ω in $\mathbb{R}^n$ such that $\Omega \supset B(x;R)$ for every $x \in K$. By adding the same constant to the operator P and the operators $P_m = A_m + q$ we may assume without loss of generality, in view of (A3.1) and (A3.2), that

$$(P\varphi,\varphi) \geq \|\varphi\|^2 \quad \text{and} \quad \int_\Omega q_-|\varphi|^2dx \leq (P\varphi,\varphi) \tag{A3.16}$$

for every $\varphi \in C_0^\infty(\Omega)$. Since $a_m^{ij}(x) \to a^{ij}(x)$ as $m \to \infty$ uniformly for $x \in \Omega$, and since we also have $[a^{ij}(x)] \geq \delta I$ for every x in Ω for some constant $\delta > 0$, it follows that there exists a sequence of positive numbers $\{\varepsilon_m\}$, with $\varepsilon_m \to 0$ as $m \to \infty$, such that

$$(1-\varepsilon_m)[a^{ij}(x)] \leq [a_m^{ij}(x)] \leq (1+\varepsilon_m)[a^{ij}(x)] \tag{A3.17}$$

for all $x \in \Omega$, $m = 1,2,\ldots.$ Combining (A3.16) and (A3.17) we find that for $\varphi \in C_0^\infty(\Omega)$:

$$\begin{aligned}(P_m\varphi,\varphi) &\leq (1+\varepsilon_m)(P\varphi,\varphi) + \varepsilon_m \int_\Omega q_-|\varphi|^2dx \\ &\leq (1+2\varepsilon_m)(P\varphi,\varphi),\end{aligned} \tag{A3.18}$$

and similarly that

$$(P_m\varphi,\varphi) \geq (1-2\varepsilon_m)(P\varphi,\varphi) \tag{A3.19}$$

From (A3.18) and (A3.19) it follows that

$$(1-2\varepsilon_m)\Lambda_R(x;P) \leq \Lambda_R(x;P_m) \leq (1+2\varepsilon_m)\Lambda_R(x;P)$$

for every $x \in K$ and $m = 1,2,\ldots,$ which implies in particular that $\Lambda_R(x;P_m) \to \Lambda_R(x;P)$ uniformly on K as claimed. ∎

Appendix 4 Proof of Lemma 5.7

To begin with we prove that if $\vartheta \in (0,1)$, $r > 2(1+2(1-\vartheta)/n)^{-1}$, $r > 1$, and $\varphi \in \mathcal{S}(\mathbb{R}^n)$, where $\mathcal{S}(\mathbb{R}^n)$ is the Schwartz space of functions of rapid decrease, then

$$\|\Lambda^{\vartheta}\varphi\|_2 \leq c(\vartheta;r)(\|\nabla\varphi\|_r + \|\varphi\|_r).$$

(Here and in the following $\|\cdot\|_r$ denotes the norm in $L^r(\mathbb{R}^n)$).

Define $B : \mathcal{S}(\mathbb{R}^n) \to \mathcal{H} = \bigoplus_{j=1}^{n} L^2(\mathbb{R}^n)$ by $(B\varphi)_j = \partial_j\varphi + n^{-1/2}\varphi$. A short calculation using the Plancherel theorem gives

$$\|B\varphi\|_{\mathcal{H}} = \|\Lambda\varphi\|_2. \tag{A4.1}$$

Note also that with $(\Lambda^{\alpha}g)_j = \Lambda^{\alpha}g_j$ we have $\Lambda^{-(1-\vartheta)}B\varphi = B\Lambda^{-(1-\vartheta)}\varphi$ so that

$$\Lambda^{-(1-\vartheta)} : \operatorname{Ran} B \to \operatorname{Ran} B \ \text{ and } \ \Lambda B^{-1}\Lambda^{-(1-\vartheta)}B\varphi = \Lambda^{\vartheta}\varphi.$$

Hence using (A4.1)

$$\|\Lambda^{\vartheta}\varphi\|_2 = \|\Lambda^{-(1-\vartheta)}B\varphi\|_{\mathcal{H}} \tag{A4.2}$$

If $f \in \mathcal{S}(\mathbb{R}^n)$ it is easy to see that

$$\Lambda^{-(1-\vartheta)}f = \Gamma\left(\frac{1-\vartheta}{2}\right)^{-1} \int_0^{\infty} e^{-t}\, t^{-(1+\vartheta)/2}(e^{t\Delta}f)\,dt$$

so that

$$\|\Lambda^{-(1-\vartheta)}f\|_2 \leq \Gamma\left(\frac{1-\vartheta}{2}\right)^{-1} \int_0^{\infty} e^{-t}\, t^{-(1+\vartheta)/2}\|e^{t\Delta}f\|_2\,dt. \tag{A4.3}$$

It is well known that $e^{t\Delta}f = K_t * f$ with $K_t(x) = (4\pi t)^{-n/2}\, e^{-|x|^2/4t}$. Using Young's inequality we have

$$\|e^{t\Delta}f\|_2 \leq \|K_t\|_p \|f\|_r$$

with $1 + \frac{1}{2} = p^{-1} + r^{-1}$. A calculation shows

$$\|K_t\|_p = c(p)\, t^{-n/2(1-p^{-1})}$$

so that using (A4.3)

$$\|\Lambda^{-(1-\vartheta)}f\|_2 \le \Gamma(\frac{1-\vartheta}{2})^{-1}\|f\|_r \int_0^\infty e^{-t}c(\rho)\, t^{-n/2(1-\rho^{-1})}t^{-(1+\vartheta)/2}dt \quad \text{(A4.4)}$$

The power of t in the integrand is greater than -1 since $r > 2(1+2(1-\vartheta)/n)^{-1}$. Thus

$$\|\Lambda^{-(1-\vartheta)}f\|_2 \le d(\vartheta,r)\|f\|_r,$$

hence from (A4.2)

$$\|\Lambda^{\vartheta}\varphi)\|_2^2 \le d(\vartheta,r)^2\sum_{j=1}^{n}\|(\partial_j\varphi+\frac{\varphi}{\sqrt{n}})\|_r^2$$

$$\le c(\vartheta,r)^2(\|\nabla\varphi\|_r + \|\varphi\|_r)^2$$

which proves (5.24).

We now turn to the proof of (5.25). It is clear that if $\vartheta \in (0,1)$ and $x \ge 0$ there exists a constant c such that

$$x^{\vartheta} \le x + c.$$

Thus for $\alpha > 0$ $(\alpha x)^{\vartheta} \le \alpha x + c$ so that

$$x^{\vartheta} \le \alpha^{1-\vartheta}x + \alpha^{-\vartheta}c.$$

Let $\varepsilon^2 = \alpha^{1-\vartheta}$. Then

$$x^{\vartheta} \le \varepsilon^2 x + \varepsilon^{-2\vartheta/(1-\vartheta)}c.$$

Since Λ^2 is a positive operator

$$\Lambda^{2\vartheta} \le \varepsilon^2\Lambda^2 + \varepsilon^{-2\vartheta(1-\vartheta)}c$$

Thus for $\varphi \in C_0^\infty(\mathbb{R}^n)$,

$$\|\Lambda^{\vartheta}\varphi\|_2^2 \le \varepsilon^2\|\nabla\varphi\|_2^2 + \varepsilon^2\|\varphi\|_2^2 + e^{-2\vartheta/(1-\vartheta)}c\,\|\varphi\|_2^2.$$

Taking square roots gives

$$\|\Lambda^{\vartheta}\varphi\|_2 \le \varepsilon\|\nabla\varphi\|_2 + c_1(\varepsilon^{-\vartheta/(1-\vartheta)}+\varepsilon)\|\varphi\|_2$$

for some constant c_1. This proves (5.25) and completes the proof. ▪

Bibliographical Comments

Chapter 1. A special case of Theorem 1.5 was proved by Lithner [25]. A version of Theorem 1.5 which holds for solutions of elliptic operators on Riemannian manifolds was described by us in [1].

Theorem 1.5 which states that solutions of certain second order elliptic equations in unbounded domains which do not grow too fast in fact decay rapidly belongs to a group of results sometimes referred to as "Phragmen-Lindelöf type theorems". Results of this general type for solutions of possibly higher order elliptic equations were studied by various authors. In this connection see P.D. Lax [24] and Agmon and Nirenberg [4; p. 220].

Chapter 3. There is an extensive literature on the self-adjointness problem for the Schrödinger operator $-\Delta+q$ and for more general elliptic operators on $\mathbb{R}^n$. Among the many papers on the subject we mention here only the older papers by Stummel [41] and by Ikebe and Kato [19], and the more recent papers by Kato [20, 21]. A detailed bibliographical commentary on the subject can be found in Reed and Simon [34]. In his more recent work Kato studied the self-adjointness problem for $-\Delta+q$ under minimal assumptions on the potential q. Thus for instance in the paper [21] no assumptions are made on the positive part of q except that it be locally integrable. Theorem 3.2 on self-adjoint realizations of $A+q$ which we prove in these lectures is of the same type, it is proved under "minimal" assumptions on q.

Formula (3.16) for the bottom of the essential spectrum of the self-adjoint realization of P (under more restrictive assumptions on the operator) is due to Persson [32].

Chapters 4 and 5. There is an extensive literature on exponential decay of eigenfunctions of Schrödinger operators. We shall mention here a few of those papers which deal the subject of exponential decay of eigenfunctions of N-body Schrödinger operators. For additional references the reader should consult [6].

We have already mentioned in the Introduction the papers by O'Conner [31], Combes and Thomas [7] and Simon [37] which prove the isotropic estimate (1) for eigenfunctions with eigenvalues below the essential spectrum. As a matter of fact these papers deal with eigenfunctions of N-body Schrödinger operators with center of mass removed. The Schrödinger operator in question is the same as the operator H studied in Theorem 4.13. It acts on $L^2(X)$ where $X \subset \mathbb{R}^{\nu N}$ is defined in (4.69). O'Conner's result is that if $\psi(x)$ is an eigenfunction of H with eigenvalue $\mu < \Sigma = \inf \sigma_{ess}(H)$, then

$$\int_X |\psi(x)|^2 e^{2\alpha|x|} dx < \infty$$

for any $\alpha < \sqrt{\Sigma - \mu}$. Here $|x|$ denotes the norm of x in the inner product (4.68). Combes and Thomas have simplified consideraby O'Conner's proof while Simon has shown that the L^2 bound could be replaced by the pointwise bound: $|\psi(x)| \le C_\alpha \exp(-\alpha|x|)$ for any $\alpha < \sqrt{\Sigma - \mu}$, C_α some constant.

The more refined non-isotropic upper bounds for eigenfunctions of N-particle systems were discussed in the literature in a paper by Deift, Hunziker, Simon and Vock [10] and in papers by Ahlrichs, M. Hoffman-Ostenhof, T. Hoffman-Ostenhof and Morgan [17,15,14] (see [6] for additional bibliography). These papers give various concrete upper bounds for eigenfunctions of multiparticle Schrödinger operators (mostly of atomic type). We also mention here the paper by Mercuriev [26] where precise asymptotic results are given for eigenfunctions of three-particle Schrödinger operators with short-range potentials.

The non-isotropic upper bounds for eigenfunctions of multiparticle Schrödinger operators which we prove in Theorem 5.2 and Theorem 5.3 were announced by us in [1] and [2]. The estimates given in these theorems turn out to be "best possible" in case the eigenfunction ψ is the *ground state* (under somewhat stronger assumptions on the potentials). This follows from similar *lower bounds* which can be established in this case. To be precise, suppose that the eigenfunction ψ considered in Theorem 5.2 or in Theorem 5.3 is the ground state which we may assume to be everywhere positive. Let $\rho(x)$ be defined as in Theorem 5.2 or Theorem 5.3, respectively. Then the following lower bound holds:

$$\psi(x) \geq c_\varepsilon e^{-(1-\varepsilon)\rho(x)}$$

for all $x \in X$ for any given $\varepsilon > 0$ and some constant $c_\varepsilon > 0$. Combining this lower bound with the upper bound established in Theorem 5.2 and Theorem 5.3, respectively, we obtain for the ground state the following asymptotic relation:

$$\lim_{|x| \to \infty} \frac{\log \psi(x)}{\rho(x)} = -1.$$

The lower bound described above for the ground state of the N-body problem (case of Theorem 5.3) is due to Carmona and Simon [6]. For a special case of Theorem 5.2 (the ground state of two electron atom) the lower bound is due to T. Hoffman-Ostenhof [16] and to Carmona and Simon [6]. (For some special lower bounds see also Bardos and Merigot [5].)

The technique used in the proof of Theorem 5.1 is essentially that of Moser [28]. The added complication in our case is due to the fact that we consider *complex* solutions u of $Au+qu = 0$ for quite general complex functions q. The special case of Theorem 5.1 when q is a real function in $L^p(B)$ with $p > n/2$ follows from the results of Stampacchia [40; see Theorem 5.1 and Remark 5.2 on p.171].

References

[1] S Agmon, *On exponential decay of solutions of second order elliptic equations in unbounded domains*, Proc. A Pleijel Conf, Uppsala, September 1979, pp 1-18

[2] S Agmon, *How do eigenfunctions decay? The case of N-body quantum systems*, Proc VIth Int. Conf Math Phys Berlin 1981, Lecture Notes in Physics, Springer-Verlag, 1982

[3] S. Agmon, *Lectures on Elliptic Boundary Value Problem*, Van Nostrand, Princeton, 1965

[4] S Agmon and L Nirenberg, *Properties of solutions of ordinary differential equations in Banach space*, Comm Pure Appl Math *16* (1963), 121-239

[5] C Bardos and M Merigot, *Asymptotic decay of the solution of a second order elliptic equation in an unbounded domain Applications to the spectral properties of a Hamiltonian*, Proc. Roy Soc Edinburgh *76 A* (1977), 323-344

[6] R Carmona and B Simon, *Pointwise bounds on eigenfunctions and wave packets in N-body quantum systems, V: Lower bounds and path integrals*, Comm Math Phys 80 (1981), 59-98

[7] J M Combes and L Thomas, *Asymptotic behavior of eigenfunctions for multiparticle Schrödinger operators*, Comm. Math Phys *34* (1973), 251-276

[8] R Courant and D Hilbert, *Methods of Mathematical Physics, Vol II* Interscience, New York, 1962

[9] E De Giorgi, *Sulla differenziabilita e l'analicita delle estremali degli integrali multipli regolari*, Mem Accad Sci. Torino Cl Sci Fis Mat Nat, Ser 3, *3* (1957), 25-43

[10] P Deift, W. Hunziker, B Simon and E Vock, *Pointwise bounds on eigenfunctions and wave packets in N-body quantum systems, IV*, Comm Math Phys. *64* (1978), 1-34

[11] P Finsler, *Über Kurven und Flächen in Allgemeinen Räumen*, Birkhäuser, Basel, 1951

[12] D Gilbarg and N S Trudinger, *Elliptic Partial Differential Equations of Second Order*, Springer-Verlag, Berlin and New York, 1977.

[13] S Helgason, Differential Geometry and Symmetric Spaces, Academic Press, New York, 1962

[14] M Hoffmann-Ostenhof and T Hoffman-Ostenhof, *"Schrödinger inequalities" and asymptotic behavior of the electron density of atoms and molecules*, Phys Rev *16 A* (1977), 1782-1785

[15] M Hoffmann-Ostenhof, T. Hoffmann-Ostenhof, R Ahlrichs and J. Morgan III, *On the exponential falloff of wave functions and electron densities*, *Mathematical Problems in Theoretical Physics*, Lecture Notes in Physics no *116* (1980), 62-67, Springer-Verlag

[16] T Hoffmann-Ostenhof, *A lower bound to the decay of ground states of two electron atoms*, Phys Letters *77 A* (1980), 140-142

[17] T Hoffmann-Ostenhof, M Hoffman -Ostenhof and R Ahlrichs, "Schrödinger inequalities" and asymptotic behavior of many-electron densities, Phys Rev *A 18* (1978), 328-334

[18] H Hopf and W Rinow, *Über den Begriff der vollständigen differentialgeometrischen Fläche*, Comment Math Helv *3* (1931), 209-225

[19] T Ikebe and T Kato, *Uniqueness of the self-adjoint extension of singular elliptic differential operators*, Arch Rational Mech Anal *9* (1962), 77-92

[20] T. Kato, Schrödinger operators with singular potentials, Israel J. Math *13* (1972), 135-148

[21] T. Kato, *A second look at the essential selfadjointness of the Schrödinger operators*, *Physical Reality and Mathematical Description*, D Reidel Publishing Co , 1974, pp 193-201

[22] S Kobayashi and K Nomizu, *Foundations of Differential Geometry*, Interscience, New York, 1963

[23] O A Ladyzhenskaya and N N. Ural'tseva, *Linear and quasilinear equations of elliptic type*, Izd Nauka, Moscow, 1964. English translation, Academic Press, New York, 1968

[24] P. D Lax, *A Phragmen-Lindelöf theorem in harmonic analysis and its application to some questions in the theory of elliptic equations*, Comm Pure Appl Math *10* (1957), 361-389

[25] L Lithner, *A theorem of the Phragmen-Lindelöf type for second order elliptic operators*, Ark for Mat *5* (1964), 281-285

[26] S P Mercuriev, On the asymptotic form of three-particle wave functions of the discrete spectrum, Sov J Nucl Phys *19* (1974), 222-229

[27] C B Morrey, Jr , *Multiple Integrals in the Calculus of Variations*, Springer-Verlag, Berlin and New York, 1966

[28] J Moser, *A new proof of De Giorgi's theorem concerning the regularity problem for elliptic differential equations*, Comm Pure Appl Math 13 *(1960) 457-468*

[29] J Nash, Continuity of solutions of parabolic and elliptic equations, Amer J Math *80* (1958), 931-954

[30] I P Natanson, *Theory of Functions of a Real Variable, vol II*, Frederick Ungar Publishing Co , New York, 1960

[31] T O'Connor, *Exponential decay of bound state wave functions*, Comm Math Phys *32* (1973), 319-340

[32] A Persson, *Bounds for the discrete part of the spectrum of a semi-bounded Schrödinger operator*, Math Scand *8* (1960), 143-153

[33] M Reed and B Simon, *Methods of Modern Mathematical Physics, I, Functional Analysis*, 2nd Edition, Academic Press, New York, 1980

[34] M Reed and B Simon, *Methods of Modern Mathematical Physics, II, Fourier Analysis, Self-Adjointness*, Academic Press New York, 1975

[35] M Reed and B Simon, Methods of Modern Mathematical Physics, *IV, Analysis of Operators, Academic Press, New York, 1978*

[36] M Schechter, *Spectra of Partial Differential Equations*, North-Holland Publishing Co , Amsterdam-London, 1971

[37] B Simon, *Pointwise bounds on eigenfunctions and wave packets in N-body quantum systems, I*, Proc Amer Math Soc *42* (1974), 395-401

[38] B Simon, *An abstract Kato's inequality for generators of positivity preserving semigroups*, Indiana Univ Math J *26* (1977) 1067-1073

[39] G Stampacchia *Le probleme de Dirichlet pour less equations elliptiques du second ordre a coefficients discontinus*, Ann Inst Fourier *15* (1965) 189-258

[40] G Stampacchia, *Equations Elliptiques du Second Ordre a Coefficients Discontinus*, Lecture Notes no 16, Mathematiques, Les Presses de L'Universite de Montreal 1966

[41] F Stummel, *Singuläre elliptische Differentialoperatoren in Hilbertschen Räumen*, Math Annalen *132* (1956), 150-176

[42] N S Trudinger, *Linear elliptic operators with measurable coefficients*, Ann Sc Norm Sup Pisa *27* (1973), 265-308

Library of Congress Cataloging in Publication Data

Agmon, Shmuel, 1922-
Lectures on exponential decay of solutions of second order elliptic equations.

(Mathematical notes ; 29)
Bibliography: p.
Includes index.
1. Differential equations, Elliptic--Numerical solutions. 2. Schrödinger operator. 3. Eigenfunctions. I. Title. II. Series: Mathematical notes (Princeton, N.J.) ; 29.
QA377.A48 1983 515.3'53 82-14978
ISBN 0-691-08318-5

Shmuel Agmon is Professor of Mathematics at The Hebrew University of Jerusalem.

Milton Keynes UK
Ingram Content Group UK Ltd.
UKHW021136080724
445247UK00005B/215

9 780691 613673